Computer Based Fax Processing

First Edition

The Complete Guide to Designing and Building Fax Application

by Maury Kauffman

The Kauffman Group
Cherry Hill, NJ

Dedication

To Hak and Adele,

founding partners

of The Kauffman Group

TABLE OF CONTENTS

Preface

In 1990, I left the frenzied world of pharmaceutical marketing to publish a newsletter, via fax. My degree in computer management told me fax broadcasting was possible, conceptually. But was it doable? I met Joe Kowalczyk, then president of Fax Interactive, one of the US's first enhanced fax service bureaus. A few weeks later, I launched OPPORTUNITIES, a marketing newsletter for hoteliers.

After three months, I realized selling advertising space wasn't for me. But, fax broadcasting was! Thus began the search for knowledge of all things enhanced fax.

And here I am, today. Quite luckily, writing a book for the undisputed heavyweight champion of all things CTI, Harry Newton.

"Writing" is not the best adjective. I am more an editor than author. In fact, I had more help with this project than Webster had with his first edition.

The names that follow are truly champions. Industry big-wigs who *really* understand what Group 1, Class 2, T.30 and X.400 mean. Not only do they appreciate this archaic, propeller-head stuff, they can form credible arguments for or against there installation or usage. I've learned much from all of them. Without their help, **Computer-Based Fax Processing** would not have been produced.

From GammaLink, a standing ovation to John Taylor, VP of Engineering, from whose brain the concept and most original thinking for this book sprang forth.

Cheers to Rich Willis, Director of Marketing, who not only supplied key parts to several chapters, but acted as my personal MIS consultant. Rich explained much of what I didn't understand and guided my progress. I couldn't have finished without his help.

The chapter on Building a 24 Port Fax Broadcasting System was written by Steve Shaw, GammaLink's Manager of Technical Services. Steve's circumvented the globe, (and Asia with me,) educating all on CBF wherever he goes. His unique ability to mix code and words, in a readable format, will take him far, (no doubt to other galaxies.)

Thanks also to Farah Siddiqui Bullara and David Taylor for their help.

Finally, at GammaLink, much thanks and appreciation to Larry Fromm, Director of Fax Products. Larry was my point man who guided me and his people and helped to keep us all on track.

From Dialogic Corp, there was Terry Henry, Market Development Manager. Before I met Terry, I thought busses were what used to take me to school. Boy was I wrong. Terry authored the chapter on Computer Telephony Busses, something I could never do. She's the designated SCSA evangelist and speaks fluent Japanese. (There's a combination.) Terry - thank you.

Unquestionably thanks to Howard Bubb, President & CEO of Dialogic Corp, who was there, whenever I needed him. Howard's open door policy is one the rest of the industry should emulate.

Special thanks also to Brigitte Davis, who alone knows the unique role she played in this project. Mucho Gracias!!

In 1988, Bob Edgar founded Parity Software. Much of my telephony and voice chapters came from him and his book, **PC-Based Voice Processing**, (which can be purchased at 1-800-LIBRARY.) I highly recommend it. I thank Bob and Laura McCabe, Director of Sales & Marketing for their advice and cooperation. I don't claim to understand exactly how their programming language VOS works, but it sure saves a great deal of time.

The section on using Brooktrout Technology's equipment couldn't have been included without Andy O'Brien, VP Marketing and Business Development's ongoing help. Andy understands markets and channels - the stuff I really like. Thanks also to Eric Giler, Brooktrout's President for his support and advice.

Though still a relatively small industry, the world of enhanced fax has many facets, and its experts, specialized in each, were there to help.

Ney Grant, President of Ibex Technologies, who trusted me before anyone else did, and to whom I will always be grateful, contributed the chapter on Building a Fax-On-Demand System. If you haven't seen his FactsLine for Windows demo - do so, it's IMPRESSIVE.

Dave Rae, President of SpectraFAX Corp, who is widely considered *the most trusted man in fax*, assisted with the overall layout and conceptual issues of the book. Thanks Dave! John Ten Eyck, SFC's Technical Writer contributed the chapter on Purchasing Turnkey Fax-On-Demand Solutions. Don't build without reading it.

At Instant Information, the Boston-based enhanced fax service bureau, hefty thanks to Mike Ryherd, Executive VP of Technical Services for helping in the beginning and filling in the pieces in the end. Mike's the only propeller-head I know with marketing sense; a rare quality. Thanks also to Brian Cavoli, Marketing Manager, for putting up with me.

And a most sincere **THANK YOU** to Joe Kowalczyk, II's President. Joe, who has forgotten more about fax and voice than I'll ever hope to know, has been more than an a mentor to be, I'm proud to call him my friend. *Joe, I couldn't have gotten here without you!*

Thanks to my family, friends and clients whom I know I've neglected, while working on this project.

Finally, **BRAVO** to the man, whom without his help, this project would not have left the gate, Harry Newton, publisher of *Teleconnect* and *Computer Telephony* magazines and editor of the Telecom Library series of books. Harry persuaded me that this project was absolutely necessary. He offered support and guidance every step of the way. I am most proud to be working with him. And, thank you to Christine Kern for answering my 1001 questions patiently and with understanding.

As a consultant, I have little more than my reputation. These are good people with excellent products and services. If you need help, look no further.

If you liked this book, or didn't...
If I've made a mistake...
If I've left something out...
If you've got an idea, I need to know about....

Please write: c/o Christine Kern
Telecom Library
12 West 21st Street
New York, NY 10010
212-691-8215, FAX: 212-691-1191

All suggestions become the property of the author. I can't pay you, but I do thank you and won't forget you.

Happy faxing!!

Maury Kauffman
Cherry Hill, NJ
Fall, 1994

Overview

Or, Why I Love This FAX Stuff!

What's:

* faster and cheaper than the Postal Service?
* less intrusive than phone call?
* easier to program than a VCR?
* connects you to millions of people without bits, bauds or parity?

Facsimile!

Nothing in the world does its job faster, easier or cheaper than fax.

In an age where photocopiers are larger than automobiles and E-Mail requires 50 page handbooks, *the joy of fax* has become one of life's simple communication pleasures.

What did you do before you were fax empowered? Can you remember, or is it blocked from memory? The effortless, elegant fax machine has revolutionized communications worldwide. (If you think everyone has a fax machine in the US, across the Orient, Japan and Hong Kong {for example} have much higher fax machine penetration rates that here.)

No one's sure exactly how it works... but it works, everywhere!

Through the close of the 20th Century, fax technology has become as much a part of our lives as have the telephone and electricity. We no longer pay attention to the marvel of sending our voice speeding through space, or wonder about the endless supply of power that pours from the sockets in our walls. We only notice these miracles when, for whatever reason, they do not work.

What was once considered merely a convenience for attorneys, (to discuss complicated documents,) has become a vital part of doing business. A company's very existence may hinge on whether a certain fax got through in a timely, error-free manner. Even mom-and-pop sandwich stores depend on fax. If this seems extreme, ask yourself when was the last time you saw a business card without a fax number? (If you can remember, you're probably also noticing how young policemen, doctors, and presidents are getting.)

Since fax has become an everyday miracle, it is taken for granted. Fax is simple to install, easy to use, and reliable. You know that when the fax machine regurgitates your document and beeps, whomever you sent information to, has it. No need for a second thought.

But the times, they are a-changing...

Just as the telephone and cable services continue to evolve and offer more services and options to meet user's changing needs, so too does fax technology.

As little as five years ago, the basic concept of faxing newsletters directly to readers wasn't even considered. Today, the US is home to dozens of enhanced fax service bureaus that send thousands of faxed pages daily. Manufacturers, VARs and System Integrators are offering Computer-based fax (CBF) systems that store and forward documents and information on demand, twenty-four hours a day, 365 days per year. Individuals can now get the information they want, when they want it, automatically, using nothing more than a phone and fax.

The familiar fax machine itself is shifting and evolving. The stand-alone box is rapidly moving into the computer or the LAN. There, it offers more options, possibilities, and intriguing ways of doing business. The pressing need to use computer-based fax technologies in the international arena is opening further exciting opportunities that can be reached by this developing fax technology.

However, before to you move beyond the familiar stand-alone fax machine, you must ask yourself many important questions: When do I use computer-based fax? Will it be available in every PC or on the LAN? Do I standardize on modestly priced systems or do I purchase the top of the line? How do I factor in possible growth into my system? Will my system be compatible with the worldwide installed base of fax devices? And many others.

It is my hope and the hope of my many contributors that this book will answer these and other questions and help you avoid the land mines you may find on your way to the future.

Section I

Introduction to Computer-Based Fax

Chapter 1

What is Computer-Based Fax (CBF) Processing?

Introduction

The term Computer-Based Fax (CBF) Processing means little more than shoving a fax board into a computer. However, what CBF Processing offers, is a great deal more.

Originally, fax boards were viewed simply as a way for standalone PCs to emulate fax machines. Thus, applications were limited to the transmission of ASCII files. Although fax transmission has always been used for highly visual data, poor quality reproduction was a fact of life until the advent of CBF.

CBF boards, like those manufactured by GammaLink and Brooktrout, enable users to receive faxes, print them and store them on a hard disk, as well as to create them using text and graphics files, and then transmit them. Depending on system configuration, these boards can act as fax servers for vast computer networks, sending, receiving, and routing faxes for entire multinational corporations.

A fax board is not a fax machine.

What is a fax machine?

A fax machine is a combination of three different devices: a scanner (used only when sending a fax), a modem (used for both sending and receiving), and a printer (used only when receiving a fax).

When you send a fax using a machine, a piece of paper is put into the machine. When the number is dialed, and the remote fax machine is reached, the sending machine scans the document at a resolution of roughly 100 x 200 dots per inch or 200 x 200 dots per inch (for more detailed information on the resolution of faxes, see Chapter 7).

The scanning process changes the image on the page into digital information. This information is then sent over the telephone line via the machines internal. The fax modem in the fax machine is similar to modems that you may buy for your computer, except that it only knows how to converse with other fax machines, not to data modems in computers. When the digital information is received by the receiving fax machine it is converted back into an image and printed.

So what is a fax board?

When using a fax board, the individual components of the machine are separated. The fax modem is taken out of the machine and put into a computer. To emulate a fax machine, the computer must be relied on to scan (or otherwise convert) documents into fax format.

On received faxes, the computer must also provide the equivalent of printing faxes, which also may be displaying them. So by now you may be asking yourself, if a computer based fax board is a fax machine minus a printer and minus a scanner, why do they sometimes cost more than an entire machine? The reason is software.

In the case of the fax machine, the only software that is needed is internal software to provide the 3 basic functions listed above. A fax board may contain a lot more computer related software to provide the glue that allows a user to convert documents into fax format, to print and display faxes, to manage the queue of outgoing faxes, to keep track of what faxes went to which numbers, to keep track of which faxes failed and for what reason, to manage more than one board in a computer at the same time, etc. In essence, once the process is computerized, a lot more information related to the faxing tasks can be tracked and controlled.

There are many other advantages to CBF Processing. They include:

Higher Quality Output
Labor Saving
Convenience
Call Progress Monitoring
Confidentiality
Efficient use of Telephone Lines
Reduced Telephone Expenses

Higher quality output

Computer based fax can produce much higher quality fax pages than the normal quality that is received when faxes are sent via a fax machine. The scanner in a fax machine is subject to inaccuracies due to the paper not being perfectly aligned in the machine and round-off error in the scanning process. With a fax board, the computer knows that the output will be printed on a fax machine at fax resolutions, so it can much more intelligently convert documents into an image that is very readable by the recipient of the fax. Some of the advanced boards on the market support conversion from advanced page description

17

languages such as PostScript and PCL. When documents are converted from these page description languages directly into fax format, the results can be stunning.

Labor Savings

The labor saving is tremendous, especially if sending the same or very similar document to more than one different person. Computer based fax is great labor saving device if any sort of repetitious faxing is done. For example, many marketing departments have data sheets and price lists on their products that they frequently send out to prospective customers. Rather than having a secretary spend a portion of each day sending the same information to different fax numbers, this can easily be done via the computer. The customer specific information such as telephone number, fax number, and address is taken one time. Once the data is in the computer, the customer can receive any documents that are stored on the computer. Another variation on this theme is to have customers call up a computer and choose among documents that have been previously stored on the computer. These systems are known as fax-on-demand systems and are described in more detail in Section VII.

Convenience

For those documents created or stored in computer systems.

This benefit is similar to the labor saving benefit above. But in this case, the reference is to any document that is created on the computer. To print out a document, then take that same document and feed it into a fax machine is a cumbersome process that can be streamlined by sending the document directly from a computer. For example, let's say that I want to send a message about an upcoming meeting to five individuals, 3 of whom are in my company and 2 who are not. I get on my email system and send an email to the 3 who are in my company. Since I have a sophisticated CBF system, I just type in the names and fax numbers of the 2 individuals who are not in the company and the same message goes to them as well.

18

This is much easier than printing out the email message and sending it by hand.

Call Progress Monitoring

Computer based fax can monitor the progress of each call and handle each fax task differently depending on the results of the first call. For example, let's say that I want to send a newsletter to 50 subscribers. On the first attempt to send the newsletter, 30 of the faxes go through the first time, 10 of the faxes encounter a busy signal, 5 get a voice on the line, 2 receive a paper out indication, on 2 there is no answer, and on the last one there is a message that the number has changed. For the busy faxes, I want to try to send again in a short amount of time, say 5 minutes. For the voice numbers, I want to not try the fax again, but alert the sender that the numbers are not fax numbers and that they should be checked out. For the 2 numbers where I get a paper out indication, I want to try again at one hour intervals, hoping that the user has replenished the paper by then. I may need to continue trying to send for several days. On the 2 that receive no answer, I will probably want to try again at a long interval (maybe the user accidentally unplugged the machine from the telephone line). If I am not successful after a predetermined time interval (say a day or two), then I should let the sender know that no one answers at these numbers. One the last number, where the number has been changed, I dont want to try sending again, but notify the sender that the number has changed, and that he should call the number to find out if there is a new number. Thus, by monitoring the progress of each call, the computer based fax system allows the sender to take much of the labor out the process of sending to machines that are not always 100% available.

Confidentiality

Receiving faxes via fax machine can compromise the confidentiality of incoming faxes in many cases because the fax machine is in a public area. Computer based fax can address this problem in several different ways. One way is to install a lan-fax server and give every recipient their own DID fax number. Each fax goes directly to the user and only the user determines when to print out his faxes or view them confidentially on his screen. Another way is to use fax mailboxes. A fax mailbox is a fax number that holds all of a user's faxes until the users is ready to print them out. The user can retrieve faxes while traveling or can go to any fax machine and retrieve them without worry about someone else seeing the faxes that were intended only for him.

Efficient use of Telephone Lines

Or, better stated, more efficient use of expensive telephone lines.

In larger organizations, many users may have the need to send or receive faxes. Once they see the benefits of computer based fax, they all may want to participate and ask for a second telephone line to be hooked up to their computer for the occasional fax that they need to send. Having a separate line for each user can be much more expensive that by concentrating the fax traffic onto a much smaller number lines (a fax server). By funneling all fax traffic through a much lower number of telephone lines, the monthly charge for basic telephone service can be reduced dramatically.

Reduced Telephone Expenses

Some of the leading fax boards on the market support the most advanced compressions and line speeds that are present in the installed base of fax machines. (For more information on fax compression, see chapter X). This can make a large difference in a long distance telephone bill if there are a lot of faxes being sent long distances, especially internationally. 30 to 50% savings are possible if fastest speeds and densest compressions are used.

Chapter 2

The History of Fax

The concept of sending electrically a reproduction of an image to a distant recipient is almost as old as the telegraph itself; in fact, it could be said they were almost born together. In those days, fax used telegraph technology to transmit: a low-speed direct current (DC) line with a single wire to ground, and no amplifiers other than slow electrical relays which curtailed the distance which these signals could be sent across. Both telegraph and fax sent their information by the same means: contact switching of metal patterns and interrupted current, provided by wet cell batteries, to send off the information. Both received by marking on paper.

The first successful fax was patented in 1843 by a Scottish inventor, Alexander Bain. His Recording Telegraph worked over a telegraph line, using electromagnetically controlled pendulums for both a driving mechanism and timing. At the sending end, a stylus swept across

23

a block of metal type, providing contact scanning wherever the type stood out from the block. This caused a voltage to be applied to a similar stylus at the receiving end, reproducing an arc of the image on a block holding a paper saturated with an electrolytic solution which discolored wherever an electric current was applied through it.

The blocks at both ends were lowered a fraction of an inch after each pendulum sweep until the image was completed. Bain's device transmitted strictly black and white images: it was unable to produce a scale of grays in between, but almost two centuries ago this was not of too vital importance, compared to the capability to send an image (regardless of how poor) over telegraph wires! Soon after Bain's invention, several versions of his idea found relatively wide application.

The first commercial fax service was started in 1865 by Giovanni Casselli, using his Pantelegraph machine, with a circuit between Paris and Lyon, which was later extended to other cities. By the 1930's, systems using photoelectric sensors and rotating drums were commonplace in newspaper offices and law enforcement agencies to send and receive photographs and other graphic material over telephone wires.

The first versions of what we would recognize as an electronic (rather than purely electromechanic) fax machine used photoelectric tubes to measure the brightness of each spot on a document's surface as it was quickly rotated on a drum. This allowed it to transmit gray-scale information. These strictly analog devices worked basically by producing a varying voltage based on the output of the photoelectric tube, with the process being reversed at the receiving end.

A problem with this system was that since there might be variations in brightness at a frequency below that of the audio range of telephone lines, the public switched telephone network could not be used. DC-coupled leased-line circuits were necessary.

Amplitude modulation (AM) was used to solve the problem of low frequencies. The varying brightness was no longer used to vary the voltage on a line but, instead, to modulate an AM carrier. Although the concept worked well, it was extremely sensitive to line noise and if circuit gain changed during transmission the image was distorted (sometimes catastrophically) by light and dark bands. Specially conditioned leased lines were still a requirement.

Other modulation schemes were tried and used: frequency modulation (FM), phase modulation (PM), and vestigial sideband (VSB) modulation, the latter a form of AM that compresses required bandwidth. After the Second World War there was great interest on the part of newspapers in using fax technology to send newspapers directly to subscribers' homes, but the coming of television as well as technical problems forced an abandonment of the idea.

We are accustomed to the machines which appeared during the course of the last decade, when scanners and thermal printers did away with spinning drums, making fax technology available to those outside of journalism, the military, and law enforcement. Businesses were the first to profit and, in increasing quantities, personal users. But before this could happen the Babel of languages these machines used, which kept them from communicating with each other, somehow had to be brought under control.

The Coming of Standards

It was not until October 1966 that the Electronic Industries Association proclaimed the first fax standard: EIA Standard RS-328, Message Facsimile Equipment for Operation on Switched Voice Facilities Using Data Communication Equipment. This Group 1 standard as it later became known, made possible the more generalized business use of fax. Although Group 1 provided compatibility between fax units outside North America, those within this region still could not communicate with other manufacturers' units or with Group 1 machines. Transmission was analog, it typically it took between four to six minutes to transmit a page, and resolution was very poor.

U.S. manufacturers continued making improvements in resolution and speed, touting the three-minute fax. However, the major manufacturers, Xerox and Graphic Sciences, still used different modulation schemes FM and AM, again, there were no standards. Then, in 1978, the now ITU-T came out with its Group 2 recommendation, which was unanimously adopted by all companies. Fax had now achieved worldwide compatibility and this, in turn, led to a more generalized use of fax machines by business and government, leading to a lowering in the price of these units.

When the Group 3 standard made its appearance in 1980, fax started well on its way to becoming the everyday tool it is now. This digital fax standard opened the door to reliable high-speed transmission over ordinary telephone lines. Coupled to the drop in the price of modems, an essential component of fax machines, the Group 3 standard made possible today's reasonably priced, familiar desk top unit.

The advantages of Group 3 are many; however, the ones that quickly come to mind are its flexibility, which has stimulated competition among manufacturers by allowing them to offer different features on their machines and still conform to the standard. The improvement in resolution has also been a factor. Similarly to the regular television set, fax clarity or resolution depends on the number of lines present in the fax: the more lines, the clearer the image is. The standard resolution of 203 lines per inch horizontally and 98 lines per inch vertically produces very acceptable copy for most purposes. The optional fine vertical resolution of 196 lines per inch improves the readability of smaller text or complex graphic material.

Another factor has been transmission speed. Group 3 fax machines are faster. After an initial 15-second handshake that is not repeated, they can send an average page of text in 30 seconds or less. Memory storage features can reduce broadcast time even more. The new machines also offer simplicity of operation, truly universal compatibility, and work over regular analog telephone lines, adapting themselves to the performance characteristics of a line by varying transmission speed downward from 9600 bps if the situation requires it.

Modulation is the process of varying some characteristics of an electrical carrier wave (CW) as the information to be transmitted on that CW varies. The three most used kinds of modulation commonly used for communications are AM, FM, and PM.

The Telecommunications Standardization Sector (TSS) is one of four permanent parts of the International Telecommunications Union (ITU), based in Geneva, Switzerland. It issues recommendations for standards applying to modems and other areas. Although it has no power of enforcement, the standards it recommends are generally accepted and adopted by industry. Until 1993, the TSS was known as the Consultative Committee for International Telephone and Telegraph (CCITT). It is now referred to as the ITU-T.

Section II

Understanding Basic Telephony

Chapter 3

The Complete Phone Call:

Off-hook to On-hook

Introduction: The Phone Line

Most domestic telephones are connected to the telephone company's nearest exchange using a cable containing two conducting wires.

The telephone company exchange is called a *Central Office* or simply *CO*. The CO is similar to a business phone system but on a much larger scale. Phone systems and other devices which can connect calls are called *switches*. Your phone is therefore connected directly to the *CO switch*. Business phone systems which function like CO switches are called *Private Branch Exchanges*, or *PBXs* (sometimes *PABX*, with an extra "A" for "Automatic").

The connection to the phone company using two wires is called a *two-wire* connection. To distinguish it from other types of connection, it may also be described as *analog*, since sound is represented by varying current rather than by digital signals. A regular business or domestic phone line without special features may also be called a *POTS* line, for *Plain Old Telephone Service*. A line to the phone company lets you connect to a number anywhere in the world through the *Public Switched Telephone Network*, or *PSTN*.

Some business phone systems (PBXs) use more than two wires to connect a phone on a desk (the *station set*) to the PBX. The extra wires are used to send signals between the station set and PBX, which can be used to implement message waiting lights, LED displays, conferencing and other features. A "standard" voice card will not support the additional wires, and uses a connection to the PBX which is like a domestic wall socket, requiring a *two wire analog station card* in the PBX. The CO provides a small DC voltage across the two wires, called (for obvious reasons) *battery*.

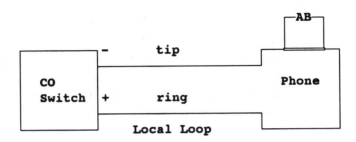

Connection of CO Switch to a Telephone

The CO provides "battery" voltage to a regular analog phone line. When the receiver is taken off-hook, the circuit is completed, "loop current" flows, and the CO switch will respond by generating dial tone on the line. When a call comes in, the CO switch applies a larger AC voltage ("ring voltage") to the line, causing the phone to ring.

The two wires are known as *tip* (often connected to battery -) and *ring* (often connected to battery +). For most purposes, it doesn't matter which way round tip and ring go, but it is wise to get it right anyway.

Starting An Outgoing Call: Getting Dial Tone

When the hand-set is taken out of its cradle, the phone is said to be *off-hook*. The action of taking the phone off-hook closes the connection AB, so that there is a complete circuit to the CO along the *local loop*. This causes current to flow, known as *loop current*. The CO switch will usually react to this by making a sound (a combination of 350Hz and 440Hz tones), known as *dial tone* which indicates that you may dial.

Most analog lines are *loop start*, which request dial tone in this way — the alternative is *ground start*, where service is requested by grounding one of the two conductors in the two-wire loop.

33

Announcing An Incoming Call: Ringing

When the hand-set is in its cradle, the phone is said to be *on-hook*. When the phone is on-hook, the connection AB is broken, but there is still a complete circuit made through a capacitor, shown as ===== in the diagram. When a call arrives, the CO applies an A/C voltage of about 105V to the circuit (*ring voltage*), and the phone rings.

Dialing A Number

There are two fundamentally different ways of dialing numbers: *tone dialing* and *pulse dialing*. Pulse dialing is sometimes called *rotary* dialing because that is the method used by old-style rotary phones.

Tone dialing uses sounds to represent digits (we include **0** through **9**, **#** and ***** as digits). Each digit is assigned a unique pair of frequencies, hence the name *Dual Tone Multi Frequency* (*DTMF*) digits, or *Touch Tone* digits. There are four *DTMF* digits (in the *fourth column* — because the usual tone pad has three) which are not usually found on telephones in the US but are used in some European countries, these are named **a**, **b**, **c** and **d**.

There are actually two standards for tone digits: DTMF and *MF* (for *Multi Frequency*). MF is used internally by the phone company, but is also used by some phone companies for *ANI* (*Automatic Number Identification*) and occasionally for other services. ANI is the business equivalent of *Caller Identification*, where the telephone number of the calling party is transmitted to the telephone receiving the call. MF is very similar to DTMF except that different pairs of frequencies are used for each digit. Some speech cards have the ability to detect MF as well as DTMF, others, especially older models, do not.

Pulse dialing uses the loop current itself to send digits. When the dial of a rotary telephone rotates, it briefly turns the switch AB on and off, thus turning the loop current on and off, resulting in "pulses" of loop current. Count one for each pulse to get the digit being dialed. You know when a digit is finished and the next one starts by the longer pause between pulses.

Pulse dialing has two major disadvantages. The first is that pulse is much slower than tone dialing, the second is that most switches will not transmit pulses over a connection. If you make a call from a rotary phone, and dial a pulse digit in the middle of the conversation, clicks will be heard at the far end, but no interruptions will be made in the loop current. This means that the only way to detect pulse digits from a remote telephone is to try to analyze the sound patterns and to "guess" when a digit has arrived. Imagine distinguishing a pulse "1" digit from the click caused by static on the line, for example — not an easy problem. Systems capable of recognizing pulse digits from the sound they make are called *pulse to tone converters*.

Completing The Call

When dialing is complete, the person or equipment which dialed the number can listen to the line to determine when and if the call is completed, ie. if the called party answers the phone. Along the way, a number of *call progress* signals may be generated to indicate how things are going in the process. Call progress signals are mostly sounds (tones) generated by a switch, some signals are made by dropping loop current briefly.

Ringing tones (called *ring-back* in the business), indicate that ring voltage is being applied to the line corresponding to the number dialed. Ring-back is generated by the CO switch which is attached to the number that you called, not by the called phone (there may be no equipment attached to the number at all).

If the dialed number is off-hook when the connection is attempted, a *busy* signal will be generated instead. If the phone company's network is busy and the local CO (the CO you are attached to) fails to make a connection to the distant CO (the CO connected to the dialed number), a *fast busy* may be generated. You don't hear fast busy tones very often — they sound similar to busy, but the pause between the beeps is shorter.

If you dial a bad number (an area code that does not exist, or a disconnected number), you will get an *operator intercept* signal (three rising tones) followed by a recording: "doo-doo-doo We are sorry, ..."

Less obvious but still important are brief drops in loop current which are sometimes generated when making long-distance calls. These can be used by the phone company as an acknowledgement that the distant CO has been reached, and are often used to indicate that the called number went off-hook. It is all too easy to confuse these brief loop current drops with the drop in loop current that signals a disconnect (end of a call due to the called party hanging up). Finally, if all goes well, the called party will answer the phone and say "hello."

Call Progress Analysis

Call progress addresses the important and difficult question: what is the state of the telephone connection?

Call progress analysis, sometimes known as *call supervision,* is most commonly performed immediately following the dialing process which communicates the desired routing of a call to a switch. The results of the attempt are indicated to the calling person or equipment originating the call in the form of *call progress tones.* Common examples include ringing tones and busy signals.

Call progress analysis presents a difficult challenge to automated equipment such as voice processing hardware because of the antiquated technology used in signaling. The most commonly used telephone connection, the analog two-wire interface, provides no separate signaling channel — all call progress signaling must be done using *in-band* signals. In other words, the only means available of signaling the progress of a call is to use the sound (audio) on the line, or the loop current. More modern digital connections such as ISDN and the European E-1 standard provide for *out-of-band* signaling, where signals completely separate from the audio channel can be used.

While some switches do provide loop current signals to indicate the completion of a call, in most cases the voice processing hardware will need to analyze the received audio on the line to match the sound against expected patterns of tones. This process is complex, and requires a great deal of tuning and flexibility if it is to be completely robust, especially when many different environments may be encountered. For example, a voice mail unit may be required to transfer a call within a PBX, dial pager units through local and long-distance calls, or forward voice mail messages over the public network to other national or international offices. It is a major challenge to provide reliable call progress features for all these different environments. Each PBX will have its own characteristic tones for ring and busy, these may even vary between different models from the same manufacturer.

A typical call progress algorithm reports the results of call progress analysis as one of the following:

Ring no answer	Ringing tones were detected, but after a pre-set number of rings, there was still no answer.
Busy	A busy tone was detected.
Fast busy	A fast busy tone, indicating that the network was unable to reach the desired CO, was detected.
Operator Intercept	The "doo-doo-doo We are sorry..." tones were detected.
No ring-back	After waiting for a set length of time, ringing tones were not detected.
Connect	The call was answered.

Call Progress Monitoring

The term call progress analysis or supervision is usually applied to the period immediately following an out-dial or transfer. A related but more general feature is *call progress monitoring*, which refers to the entire length of the call. The simplest, and generally most reliable, aspect of call progress monitoring is hang-up, or disconnect, supervision.

In analog environments, hang-up supervision is usually implemented by dropping loop current for a brief period. Digital environments will generally signal a disconnect by changing the value of a signaling bit. On standard T-1 circuits, for example, the A bit transmitted to a device will generally be changed from $A=1$ (corresponding to loop current active) to $A=0$ (corresponding to loop current absent).

The bad news is that some switches, including unfortunately many business phone systems, do not provide loop current disconnect supervision. To detect a disconnect in such an environment, it may be necessary to monitor the line continuously for the tone generated by the switch when the call has been terminated and the line is left on hook — generally a new dial tone, or a re-order tone (beeps, possibly followed by a message like "Please hang up and try dialing your number again").

The Flash-Hook

When a call is in progress, a service from a CO, PBX or other switch may be requested by making a *flash-hook*. A flash-hook puts the phone on-hook briefly — long enough for the switch to detect it, but not long enough to disconnect the call. You will probably be familiar with a flash-hook from the domestic Call Waiting feature.

On a business phone system, a flash-hook will generally give you a second dial tone, allowing you to make a three-way conference (you stay on the line, flash-hook a second time to complete the conference), or a transfer (you hang up when the second number answers).

Most local phone companies offer *Centrex* features, which allow transfers and three-way conferencing using the same flash-hook and dial sequence as a typical small business phone system.

Terminating A Call: Disconnect

To terminate a call, one end goes on-hook, ie. hangs up the phone. Sooner or later (there may be a delay of twenty seconds or more, less for a local call), the phone company will transmit the disconnect by dropping the loop current. Some phone systems don't transmit a disconnect at all, so you may have no way of detecting the end of the call at the local end.

This chapter was adapted from Bob Edgar's book: **PC-Based Voice Processing,** available from Flatiron Press, (800-LIBRARY.) I thank Bob, founder of Parity Software Development Corporation for his help and advice with this chapter. I recommend his book wholeheartedly.

Chapter 4

Call Routing

Introduction

There are various ways inbound telephone calls can be routed. Description of several basic methods follow.

DNIS And DID Services

Many phone companies offer a service which allows you to terminate several different phone numbers on the same set of lines. When an in-coming call arrives, it searches (*hunts*) for an available line and generates a ring signal. Your equipment responds with a *wink* signal, which causes the phone company switch to send you some or all of the digits that the caller dialed, usually as DTMF digits. This is called Dialed Number Identification Service, or *DNIS* (pronounced "dee-niss"). A T-1 wink is a brief off-hook period, which will usually be signaled using the A bit on a T-1 channel.

DNIS is used by answering services and service bureaus, which may be answering calls from many more numbers than they have operators or IVR lines.

An analog wink is done by reversing battery voltage. Analog DNIS is called *DID*, for *Direct Inward Dial*, and sometimes *DDI*. A typical use of DID is to provide direct numbers to employees in a large company. The company phone system obtains the last four digits of the dialed number via DID and routes the call to the appropriate extension.

To get DNIS or DID digits and answer a call, you would use a sequence like this:

1. Wait for ring
2. Send wink
3. Get DNIS digits
4. Go off-hook

T-1 Digital Trunks

A T-1 digital trunk carries 24 telephone connections on two *twisted pair* (two-wire) cables. The set of 24 connections is called a *T-1 span*. Each of the 24 connections is referred to as a *time-slot*. Each time-slot carries sound, digitized at 64Kbps, and two *signaling bits*, referred to as the *A and B bits*, which play a role similar to loop current signaling on analog lines. One time-slot is sometimes referred to as a *DS-0*, for Digital Signal level 0, signal.

While use of the A and B bits is not the same in all equipment, a common convention, called *E&M* signaling, keeps the A and B bits equal, and uses the A bit to indicate whether or not a connection is active. Thus, A bit high (set to 1) corresponds to loop current flowing, A bit low (set to 0) corresponds to no loop current.

Digits can be dialed on a time slot using the same methods as analog lines: DTMF, MF and pulse. Pulse digits are sent by turning the A and B bits on and off, just as rotary pulse dialing turns loop current on and off on an analog line. With E&M signaling, a pulse would be sent by briefly changing the off-hook state (AB = 11) to on-hook (AB = 00) and back again.

The complete T-1 span with 24 DS-0 signals and synchronization information is called a *DS-1* signal.

Automatic Number Identification (ANI)

ANI (pronounced "Annie") delivers the phone number of the person (or machine) calling. It functions in a similar way to DNIS: digits are transmitted at the start of the call, with ANI the digits are the number of the originating telephone line. These digits can be used by a VRU to block a call, route it to the responsible agent or do a database lookup to retrieve the caller's account information. There are several standards for the way phone companies deliver ANI digits. Local phone companies, like New York Telephone, deliver ANI digits in between the first and second ring. This service is known as *caller ID*. They sell it as a service for their residential subscribers to ward off harassing calls, but caller ID also has business applications.

A more useful ANI service is delivered by the long distance companies, such as AT&T, MCI, and Sprint, through their 800 lines. If you have an 800 number and you have it directly hooked up to your office on a T-1 trunk then you can receive the phone number on all the 800 calls you receive. This service is now so widespread that, with an 800 line from virtually any long distance company (not just the three I mentioned above), you can receive the phone numbers of over 95% of all the phones in the U.S. The ones you can not get are typically in outlying rural areas. How you will receive ANI digits on your 800 lines depends on your chosen long distance carrier. There are basically two ways to receive ANI from a long distance carrier — in-band or out-of-band. *In-band* means you receive the digits in the voice data link of the conversation. *Out-of-band* means you receive the ANI digits on another circuit, typically in a communications channel just devoted to signaling. At the time of writing, AT&T, for example, only provides ANI on *PRI* ISDN trunks using out-of-band signaling. ISDN stands for Integrated Services Digital Network. PRI stands for *Primary Rate Interface*. PRI ISDN trunks are an extension of today's common T-1 digital standard. But they are not the same as today's T-1.

MCI will provide ANI services in-band and out-of-band, on T-1 or directly on an individual 800 line. When you buy several 800 lines, it's less expensive to connect from your office to your chosen long distance carrier on a T-1 line.

This chapter was adapted from Bob Edgar's book: **PC-Based Voice Processing**, available from Flatiron Press, 800-LIBRARY. I thank Bob, founder of Parity Software Development Corp for his help and advice with this chapter.

Section III

Understanding Facsimile Technology

Chapter 5

The Fax Image

Resolution

A fax image is made up of dots. The dots cover the page and when combined, compose letters, numbers, pictures, etc. There are two different resolutions widely used by fax machines and fax boards. These resolutions are referred to as standard and fine. Most fax machines have some way of switching between the two resolutions. The default is usually standard.

Standard resolution is 204 x 98 dpi or 204 dots per inch in the X (horizontal) direction and 98 dots per inch in the Y (vertical) direction. The dots are not square, but rectangular. Fine mode doubles the Y resolution of standard mode; thus it is 204 x 192 dpi. Sometimes the numbers are rounded off, so people may refer to a standard mode fax to be at 200 x 100 dpi and fine mode to be 200 x 200. This gives a good comparison to other devices that you may be familiar with such as ink jet printers and laser printers of several years ago which had 300 x 300 dpi resolution, or the newer laser printers which have 600 x 600 dpi resolution. The more dpi, the clearer the image on the paper. However, the more dpi, the bigger the file, which is very important when considering fax communications since this file needs to be sent across telephone lines. The bigger the file, the longer it takes to send the image. Fine mode images take roughly twice as long to send as standard mode images.

Compression

When facsimile systems first became popular, with the establishment of Group 1 and Group 2 standards, they handled information in real time as an analog signal, in which variations in the voltage on the line were an analog representation of the intensity of the signal. The idea of processing the signal by clocking, quantizing, and storing it as digital information seemed like something that would be found aboard the starship Enterprise and not on an office desk. In those old analog days scanning and recording systems were mechanical. It was difficult to even mark a straight line down a received page without noticeable jitter (a saw-toothed effect in the printout).

Keeping the scanning and recording spots in exact step with the clock was impossible. There was also a drift between the clocks at the transmitter and receiver units. Although manual adjustment of the frequency kept the skew within acceptable limits, this was far from being an ideal solution. There existed no digital modem for sending signals and providing a locked-in sampling clock at the receiver. Practical means for the digital storage of pil information for one or two scanning lines were yet to be developed.

As digital technologies evolved, it finally became practical to use compression encoding techniques to remove redundancy (repetitious information such as white space), from the page being scanned and restore pil information for recording at the receiving end.

Compression is defined as the ability to change the form data into a representation expressed by a minimum of bits, reducing transmission time. Encoding schemes reduce sending errors and further speed broadcast times. This technique has been viewed as a leading factor in the growth of facsimile technology, transforming it from a curiosity of limited use into a ubiquitous utility, very much like indoor plumbing and electricity.

For example: If a piece of paper is 8.5 inches wide by 11 inches long and each square inch contains roughly 200 X 100 dots (in standard mode), then there are 8.5 X 11 X 200 x 100 dots for a single page. That's a lot of dots. 1.87 million to be precise. If we were to represent each dot as a bit, then a single page would take up 1.87 million /8 or 234 Kbytes (8 bits to a byte). At normal fax speeds of 9600 bits per second, a file this big would take about 6 minutes to send (trust me on the math). But most of know that a single page does not take this long to fax. How is this done?

Rather than send a bit for each potential dot on the page, the file is compressed by taking advantage of the fact that most fax pages have big chunks of space where the color does not change. The standards committees that govern fax communications used a method for compressing such data that is called Modified Huffman, sometimes referred to as MH. It is also referred to as Group 3 compression (after those fax machine that use it) and one-dimensional compression, which refers to the fact that the compression deals with each line of dots one at a time and uses no knowledge of previous lines to build subsequent lines. Another name is T.4 compression. T.4 is the name of one of the international standards specifications that defines formats for images. Modified Huffman compression will typically shrink the size of an image down to 20 to 30 Kbytes per page. Thus the time to send one page is about 20 to 30 seconds, more in line with what you are probably used to if you have watching fax machines send or receive faxes.

Note that this example is a standard mode example. The same process occurs on fine mode faxes, but they will still be about twice as large as the standard mode faxes.

After the standards committees agreed on the Modified Huffman standard, they later agreed on two other compression types that are similar in concept to Modified Huffman but use a different algorithm and typically compress images even more densely than Modified Huffman. The first is called Modified READ (MR) or Group 3, 2-D (for 2 dimensional). The difference in size between an MH compressed file and an MR compression file is roughly 15%. This percentage varies depending on the composition of the image. The other compression type is Modified Modified READ (MMR), also known as Group 4 compression (since it is the compression type specified for Group 4 fax machines). The standards committees refer to the T.6 specification for the definition of this image type. An MMR image is roughly 25% to 30% more dense than an MH image.

Not all fax machines and boards support the advanced compressions (MR & MMR), but all support MH. Most machines support MR, and about 20% support MMR. Generally, only the intelligent fax boards support either MR or MMR. The amount of money that can be saved on one's telephone bill by using the advanced compressions can be significant if there is a lot of fax traffic that is going long distance or internationally. Since the files may be up to 30% smaller, the telephone bill can be reduced by up to 30%.

How Compression Occurs

In its Recommendation T.4, the ITU-T specifies that Group 3 fax units must incorporate MH run-length encoding of scanning lines. The technique used is called run length encoding. The algorithm begins scanning at the upper left-hand corner of the image, and the image is surveyed at a specific resolution value, typically 203 dpi, a line at a time, looking for consecutive dots; that is, how many of these dots (or pils) of black or white there are before there is a transition to the opposite color. All facsimile encoding techniques operate by scanning one line at a time and then going on to the next. This is best visualized by considering the lines across a pad of notebook paper. These would be the scan lines the algorithm processes. Scanning takes place from left to right. The survey algorithm begins by determining the color of the first pil (black or white), and then counts how many dots are that color: 5, 23, or maybe the entire line, 1728 bits.

Consequently, there is no direct storage of black and white pils, only the run lengths themselves are stored. For example, the compression algorithm may determine that there are 23 white pils. This information is not transmitted as the 23 zero bits or even as a byte representing the number 23, but rather as a much shorter encoded version—a code word—of that number: 11 bits that represent the number 23. These code words have been standardized and embodied into tables, and are well-established. Thus, instead of sending a white line across a page as 1728 bits, MH sends a 9-bit code word. This compresses the information 192 times. Different code words are used to represent shorter white (or black) runs.

A total of 1728 code words would be needed to cover all possible run lengths across a page. To produce a shorter code table, the runs were grouped in multiples of 64 words in a make-up code table, with shorter runs of 0 to 63 placed in a separate, terminating code table. This reduced the matter down to only 92 binary codes for any white run lengths of 0 to 1728 pils by sending a make-up code followed by a terminating code. Black runs also have 92 different codes. These code words were chosen by looking at varied page samples and minimizing the numbers of bits needed to send each one.

Compression Techniques

Although all compression techniques operate by surveying one scan line, then proceeding on to the next, there are important differences between them.

Modified Huffman compresses only one scan line at a time; it looks at each scan line as if it were the only one. Each line is viewed by it as a separate, unique event, without remembering anything about previous scan lines. Because of this single-line characteristic, MH is referred to as a one-dimensional compression technique.

Modified READ on the other hand, uses the previous line as a reference, since most information on a page has a high degree of vertical correlation "vertical correlation"; in other words, an image (whether it is a letter or an illustration) has a continuity up and down, as well as from side to side, which can be used as a reference. This allows MR to work with only the differences—the variable increments (rates of change), or deltas—between one line and the next. This results in a rate of about 35 percent higher compression than is possible with MH. MR operation may at first appear as a complex concept to grasp, but it really is not. Imagine, for example, a white sheet of paper with a black circle in its middle. The sheet, being all white can be effectively compressed because, after the first compression, there is no change in the following lines so the MR algorithm just repeats the line.

When MR starts scanning the circle, black run lengths begin and, as it proceeds downward, the circle scan gets increasingly larger. However, since MR uses previous lines as a reference, instead of counting all the black pils to encode them as the MH technique would, the MR algorithm only notes the deltas — or rate of change, between lines. Therefore, the difference between MH and MR is, essentially, that the latter uses a knowledge of previous lines to reference its vertical compression. Because MR works vertically as well as horizontally, it is called a two-dimensional compression technique.

As the name implies, MMR is very similar to MR. The difference lies in how it handles errors.

When a fax signal is distorted by a noise pulse induced by electrical interference, or whatever reason, errors occur in bits transmitted over the phone line making impossible the recovery of the original pattern of black and white pils in the received copy. To prevent an incorrect pattern from propagating down the page and corrupting the entire message, a form of error correction mode (ECM) was needed.

Originally, facsimile technology had no protocol for a recovery method in cases when bits or code words were lost or garbled during transmission over the telephone network. The original Group 3 compression techniques did not have an ECM, and therefore a means was arrived at through which to resynchronize the printer when an error took place; otherwise, the entire fax could be lost. This is why both MH and MR contain a special signal called an end of line (EOL) code. If the printer receiving the fax runs into an error, it waits until receiving an EOL code in the stream, to restart its decompression procedure and resynchronize. That way, if data errors occur, the entire message is not lost. Often, rather than producing a white line, the printer will simply repeat the previous line. Generally this is not immediately obvious to the naked eye, particularly in text transmissions. In any case, although the fax may not end up looking particularly attractive, only a few scan lines have been lost, not the whole thing.

On the downside, the EOL code adds an additional 12 to 24 bits of data at the end of each scan line, lengthening transmission time.

One of the differences between MR and Modified Modified READ, is that MMR does not use the EOL code because it was designed originally for Group 4 facsimile work. Group 4 only operates on a digital network, which is inherently error-recovered because if an error occurs, the data packets are called for again and repeated. Since there is no data loss, there is no need to resynchronize the image and this results in a 12- to 24-bit-per-scan-line compression gain. Therefore, statistically, MMR provides the highest compression for normal images.

One of the important facets of compression techniques is how often a reference—versus a delta—line is sent. This is known as the K factor. With standard mode fax under MR, K is equal to two (K=2); that is, every second scan line is a reference line and then a delta line is sent. At fine resolution K=4, which means that every fourth line is a reference line. MMR's K factor equals infinity because MMR uses no MH coded lines. MMR sends a reference line at the beginning of a page, then deltas all the way down. An error-free signal is needed for this to work. If this were done in MR, without error checking, an error on any one line would propagate downward, worsening with every additional line, corrupting the information on the balance of the page.

In 1992, the Group 3 Study Group of the ITU-T allowed Group 4 (T6 encoding) compression—a variation of MMR—to become an optional transmission technique for Group 3. Since MMR can only be transmitted by a Group 3 device supporting "ECM", both the transmitting and receiving units must support ECM to work with that type of data stream.

It is interesting to note that there are times when any of these compression techniques can go the other way and generate a data stream considerably larger than the actual pil-per-pil representation. This sometimes occurs during the processing of some grey scale images (such as photographs, or desktop publishing generated documents where grey is used) where there are numerous changes, black and white, horizontally and vertically, requiring the transmission of many additional code words. This is why computer-generated grey is very slow to transmit.

Compression, Resolution, and Speed

For all intents and purposes, compression and resolution are unrelated. MH, MR, and MMR are strictly loss-less compression techniques—the compressor and decompressor algorithms that run are solely concerned with the number of pils to survey in any given dimension—therefore none of these techniques has any effect whatsoever on resolution.

This is also why speed is not a factor in compression. Compression techniques are applied to bit-map images and are unaffected by scanner or modem speeds. The evidence for this is that it is possible to have a bit-map on a computer's hard disk, apply compression techniques to it, and obtain a compressed image, all of it without the involvement of transmission or facsimile technologies.

Anyone attempting to estimate and plan for a company's fax expenses needs to understand that since different compression techniques result in different file sizes and storage requirements. An efficiently compressed file is smaller and requires less transmission time. For a user doing multipage broadcasts products that support MMR, the highest compressed image, through optional Group 3 parameters, can be a matter of vital importance to the bottom line.

The use of faster modems, transmitting at 14.4 kbps, offers the possibility of sending more bits per unit of time over the telephone. But this is solely a matter of broadcasting data faster across the wire—the level of compression (whether MH, MR, or MMR are used) remains the same. This is purely a matter of raw transmission speed, completely unrelated to image, scanner, or compression. If we think of the compressed data as a letter, then it becomes obvious that it could be sent either by train (slow modem) or plane (fast modem). Although the trip would take less by plane, whether sent by land or air it would still be the same letter. Also, transmission rate is not dependent on the sending machine, but on the receiving machine.

A traditional fax machine is a mechanical device. It must reset its scanner, and advance the page as it prints each scan line it receives. Today's machines generally have a 10-millisecond scanning line time requirement.

When fax devices warble lovingly at each other before transmission, during the handshake period, they exchange information on scanning capabilities. If the receiving machine's rate is 20 ms per line and the transmitting machine sends data at a faster rate, it will add fill bits (also called zero fill). These extra bits pad out the amount of send time, giving the remote machine the additional time it needs to reset prior to receiving the next scan line.

The amount of fill bits added by the sending machine is determined by the receiving machine's capability. If a machine that sends at high speeds broadcasts to a slower machine, some of the time-saving benefits of sending encoded data and higher modem speed may be lost if a lot of fill must be introduced.

The rate at which a machine can receive is not related to the rate at which it can send. Although CBF generally does not require fill bits, if a CBF device is sending to a traditional fax machine it must tailor its sending mode to that device's receiving capabilities.

It is important to bear in mind that because facsimile is a compressed technique, in the case of CBF, it is essential to have software-based tools that will enable the user to correctly decompress the image back into an accurate, readable form. CBF requires a knowledge of compression techniques and their compression ratios, that is, the number of bytes it takes to store images, in order to accurately estimate archiving capacity. File sizes obviously are a factor in how long a telephone call will take—the latter being directly related to cost.

Getting Hard Copy

When the receiver modem decodes the received analog fax signal regenerates the digital signal sent by the fax transmitter, the MH/MR/MMR block then expands this fax data to black-white pil information for printing.

Overall, there are two ways to convert a fax into hard copy: through a thermal printer, or a regular printer (the latter covering everything from pin to the preferred laser printer formats).

Each inch of a thermal printer's print head is equipped with 203 wires touching the temperature-sensitive recording paper. Heat is generated in a small high-resistance spot on each wire when high current for black marking is passed through it. To mark a black spot, the wire heats from non-marking temperature (white) to marking temperature (black) and back, in milliseconds.

Since thermal paper comes in rolls, a thermal printout has the advantage of adapting itself to the length of the copy being transmitted: if the person on the other end is sending a spreadsheet, the printout will be the same size as the original. With sheet-fed printers, such as a laser printer, the copy is truncated over as many separate sheets as necessary to download the information. That, however, is about the only major advantage of thermal printouts. In general, thermal printouts offer less definition, thermal paper is inconvenient to handle due to its tendency to curl, and if the copy is to be preserved, it must be photocopied because thermal paper printouts tend to fade over time. Then, most important of all, there is the matter of cost: thermal paper printouts are between five to six times more expensive than those made on plain paper. Plain paper fax machines using laser printer technology are becoming increasingly popular, particularly in the U.S.

TIFF and PCX

When a fax machine sends or receives an image, it goes from paper to paper, being converted into an image as described above. When a computer sends or receives a fax, it stores the image as a file or series of files. Two of the common file formats for storing fax images are TIFF (Tag Image File Format) and PCX (PC Paintbrush format). Both of these formats are widely used by other computer programs, such as drawing programs, graphics programs, scanning programs, etc. The fact that standard formats are used gives the user a lot of flexibility in manipulating fax images, combining them with other images, adding them into word processing documents, etc.

TIFF files of the same image typically take up about half the space of PCX files. So if you plan on setting up a system that will either contain a lot of files or receive a lot of files, you might want to remember that systems that support TIFF will require about half the disk space of systems that use PCX. Average image sizes will range from 20 - 40K per page for TIFF files and 40 - 80K per page for PCX files. The amount of space required is dependent primarily on the resolution of the page and the composition of the image. Pages containing text will usually be much smaller than pages containing pictures or drawings with large areas of gray.

Chapter 6

The Complete Fax Call: T.30

Dial to Acknowledgement

What takes place before, during and after a fax transmission may seem to fall into the realm of magic, but is in reality a very carefully orchestrated procedure. With small variations, a modern fax machine call is made up of five definite stages as defined by the ITU-T's T.30 specification: Phase A, the call establishment; Phase B, the pre-message procedure; Phases C1 to C2, consisting of in-message procedure and message transmission; Phase D, the post-message procedure; and Phase E, call release.

Phase A

Phase A takes place when the transmitting and receiving units connect over the telephone line, recognizing one another as fax machines. It is here that the handshaking procedure begins between the transmitting and receiving units. This is accomplished through a piercing, 1100-Hz tone (the calling tone or CNG), sent by the caller machine. Virtually everybody is well-acquainted with this mating call, having heard it either by dialing a fax number by mistake, or waiting to hear the answering love warble of the machine at the other end before activating your own to transmit the message.

It is here that the caller machine sends its Called Station Identification (CED) and the answerer replies with its own, a shrill 2100-Hz tone. Once this has been accomplished, both machines move on to the next step.

Phase B

In Phase B, the pre-message procedure, the answering machine identifies itself, describing its capabilities in a burst of digital information packed in frames conforming to the High-Level Data-Link Control (HDLC) standard. Always present is a Digital Identification Signal (DIS) frame describing the machine's standard TSS features and perhaps two others: a Non-Standard Facilities (NSF) frame which informs the caller machine about its vendor-specific features, and a Called Subscriber Identification (CSI) frame. At this point, optional Group 2 and Group 1 signals are also emitted, just in case the machine at the other end is an older Group 1 or Group 2 device unable to recognize digital signals. This is very rare, however, and chances are that both devices will be Group 3 machines, in which case they will proceed from the identification section of Phase B to the command section.

In the command section of Phase B, the caller responds to the answerer's initial burst of information, with information about itself. Previous to transmitting, the caller will send a Digital Command Signal

(DCS) informing the answerer how to receive the fax by giving information on modem speed, image width, image encoding, and page length. It may also send a Transmitter Subscriber Information (TSI) frame with its phone number and a Non-Standard facilities Setup (NSS) command in response to an NSF frame. Then the sender activates its modem which, depending on line quality and capabilities of the machines may use either V.27 (PSK) modulation to send at a rate of 2400 or 4800 bps, or V.29 (QAM) modulation to transmit at 7200 or 9600 bps.

A series of signals known as a training sequence is sent to let the receiver adjust to line conditions, followed by a Training Check Frame (TCF). If the receiver receives the TCF successfully, it uses a V.21 modem signal to send a Confirmation to Receive (CFR) frame; otherwise it sends a failure-to-train signal (FTT) and the sender replies with a new DCS frame requesting a lower transmission rate.

Phase C

Phase C is the fax transmission portion of the operation. This step consists of two parts: C1 and C2, which take place simultaneously. Phase C1 deals with synchronization, line monitoring, and problem detection. Phase C2 includes data transmission.

Since a receiver may range from a slow thermal printing unit needing a minimum amount of time to advance the paper, to a computer capable of receiving the data stream as fast as it can be transmitted, the data is paced according to the receiver's processing capabilities. An Error Correction Mode (ECM) procedure encapsulates data within HDLC frames, providing the receiver with the capability to check for, and request the retransmission of garbled data. Because of its encapsulation within the HDLC frames this on-going procedure does not lengthen transmission time.

Phase D

Once a page has been transmitted, Phase D begins. Both the sender and receiver revert to using HDLC packets as during phase B. If the sender has further pages to transmit, it sends a frame called the Multi-Page Signal (MPS) and the receiver answers with a Message Confirmation Frame (MCF) and Phase C begins all over again for the following page. After the last page is sent, the sender transmits either an End Of Message (EOM) frame to indicate there is nothing further to send, or an End Of Procedure (EOP) frame to show it is ready to end the call, and waits for confirmation from the receiver.

Phase E

Once the call is done, Phase E, the call release part, begins. The side that transmitted last sends a Disconnect (DCN) frame and hangs up without awaiting a response.

Chapter 7

Facsimile Standards

Introduction

As we've seen, the beginning of facsimile technology is traceable to the 1840's, when the first successful fax device was patented; and, commercial fax service began in France, in 1865.

After the Second World War there was great interest on the part of publishers in using fax technology to send newspapers directly to subscribers homes, but the coming of television as well as technical problems forced an abandonment of the idea.

Over the 1950's and 1960's, fax technology began to evolve into the form we recognize today. As fax became more widely used, it soon became obvious that standards would be needed to enable different fax machines to communicate with each other.

It was not until October 1966 that the Electronic Industries Association proclaimed the first fax standard: EIA Standard RS-328, Message Facsimile Equipment for Operation on Switched Voice Facilities Using Data Communication Equipment. This Group 1 standard as it later became known, made possible the more generalized business use of fax.

Although Group 1 provided compatibility between fax units outside North America, those within still could not communicate with other manufacturers units or with Group 1 machines. Transmission was analog, typically it took between four to six minutes to transmit a page, and resolution was very poor.

U.S. manufacturers continued making improvements in resolution and speed, touting the three-minute fax. However, the major manufacturers still used different modulation schemes FM and AM. So, again, there were no standards.

Later in 1978, the ITU-T (then referred to as the ITU-TSS) came out with its Group 2 recommendation, which was unanimously adopted by all companies. Fax had now achieved worldwide compatibility and this, in turn, led to a more generalized use of fax machines by business and government, leading to a lowering in the price of these units.

When the Group 3 standard made its appearance in 1980, fax started well on its way to becoming the everyday tool it is now. This digital fax standard opened the door to reliable high-speed transmission over ordinary telephone lines. Coupled to the drop in the price of modems, an essential component of fax machines the Group 3 standard made possible today's reasonably priced, familiar desk top unit.

The advantages of Group 3 are many; however, the one that quickly comes to mind is its flexibility, which has stimulated competition among manufacturers by allowing them to offer different features on their machines and still conform to the standard. The improvement in resolution has also been a factor. The standard resolution of 203 lines per inch horizontally and 98 lines per inch vertically produces very acceptable copy for most purposes. The optional fine vertical resolution of 196 lines per inch improves the readability of smaller text or complex graphic material.

Group 3 fax machines are faster. After an initial 15-second handshake that is not repeated, they can send an average page of text in 30 seconds or less.

Memory storage features can reduce broadcast time even more. The new machines also offer simplicity of operation, truly universal compatibility, and work over regular analog telephone lines, adapting themselves to the performance characteristics of a line by varying transmission speed from perhaps as much as 14.4 kbps, all the way down to 2,400 bps if the situation requires it.

Section IV

Computer-Based Facsimile

Chapter 8

Fax Routing

Introduction

The Group 3 protocol (the digital fax standard that determines a fax machine's capabilities) does not stipulate any kind of an auxiliary address for the inward routing of faxes. It was not until the appearance of more complex systems such as fax machines designed to keep faxes internally, in secure or private mailboxes, that a routing mechanism of sorts first became necessary. With the coming of CBF and its inherent flexibility, the possibilities opened by various routing schemes became too valuable to ignore.

The widespread use of LANs has made dependable inward routing (the process of selecting the correct destination for a message within an organization), a necessity. There are now several ways to route messages. Which one is selected and how it is implemented, has a direct effect on network fax capabilities. When studying routing schemes, a prime consideration must be the installed base of Group 3 fax machines, estimated at 30 million. If they cannot readily use the routing solution, it has missed the mark. The next most important concern must be the outside sender.

A basic consideration in routing system selection is what the sender of the fax must do for it to work. The different ways people send faxes preclude the wide use of some of the more elaborate sender-dependent routing mechanisms.

Methods requiring a trained sender generally serve a limited purpose. If a company needs to have an inward routing method used by a closed user community within its organization for a specialized task, then it can train personnel, use specialized fax machines, and create something unusual in the way of a routing scheme. The solution then meets a special need while posing no problems because the company controls the community of senders and receivers whose efficiency is increased by the system. However, in more widespread applications where the volume of faxes received is in the tens, hundreds, or even thousands of faxes a day, coming from all kinds of machines and senders, another solution is advisable.

Manual Routing

In its most basic form, manual routing consists of someone picking up incoming faxes from a regular fax machine and distributing them to the intended recipients. With CBF, manual routing uses a software utility to electronically display to an operator a listing of faxes received. Usually the information displayed comes from the cover page as well as the transaction record that shows the date and time the fax was received, how many pages it has, and how long it took to receive. The operator doing the routing scans this and can probably determine from this information whom the fax is intended for. Sometimes, however, this is not enough and the fax itself must be viewed before routing; this may be unsatisfactory when dealing with proprietary or confidential information.

DTMF Routing

Unattended DTMF routing uses the buttons on a TouchTone phone. A concern with DTMF routing is that it depends on the sender to work: anyone attempting to send in a fax must be trained in the particular routing system or somehow get operating instructions.

With DTMF routing the sender not only needs a telephone number, but an extension number as well. This may pose a problem. How is that additional set of numbers used? When the sender dials, he may expect an operator to come on. If there is no operator, suddenly there is a tone. Is the extension number entered now? What happens if a digit is missed? Then, if the sender takes too long to decide, the machine on the other end may time out, forcing the caller to start all over again.

Often, when the sender hears a sound different from the regular dial tone, the tendency is to punch the start button on the fax machine. Automated voice prompting helps unless dealing with a fax machine at the sending end instead of a human ready to enter additional number sequences when told to. Generally, if a second sequence of numbers is not entered, the fax is diverted to a default central directory for subsequent manual routing.

Direct Inward Dialing

DID is generally viewed as the best routing choice. This is because, unlike DTMF, it is the telephone company's central office (CO) not the end user that provides the routing information. All the sender does is dial a single telephone number, and the fax goes directly to the recipient's workstation. DID is the most foolproof, transparent way of routing.

DID consists of one or more trunk lines between the CO and a customer's premises. DID trunks are normally (but not in CBF's case) used to reduce the number of channels between the PBX and the CO: they are one-way trunks. A PBX perceives the DID trunk as one of its single-line phones and can interpret four-digit dialing.

Therefore, some switching functions usually done at the telephone company's CO are done at the customer's premises. Generally, each CBF board can have only one DID line, but an unlimited number of phone numbers can be installed on that line since the board hardware and software direct the incoming call from the CO.

When outside callers dial a station on a DID-configured PBX, the CO switch signals the PBX for service. An available trunk is activated by the CO by causing a current to flow. The PBX detects the current flow but, rather than responding with a dial tone (which it would have done if it really considered this loop to be a true station-to-PBX loop), it gives the CO a wink-start signal. The wink informs the CO that the PBX is ready to receive the incoming addressing information. The wink signals the CO switch to send the identifying digits dialed by the distant user, which the PBX uses to complete the connection. These last digits provide the data that the system uses for inward routing. After the pulses have been received by the PBX and the desired station has answered the call, the PBX signals the CO that the call and billing can begin, by reversing the polarity of the current as it did for the wink-start signal, but leaving it reversed for the entire duration of the call. When the call is finally disconnected, the DID facility returns to its idle state.

DID installation requires an awareness of polarity and power requirements. With standard conventional telephone facilities this is not a consideration, but with DID it is all-important. Any business considering the use of DID should use its in-house resources and invite the participation of its MIS, computer, and telecommunications experts in planning for the company's network fax system. Once the DID system is up and working, no further attention is needed and it will operate transparently for both sender and recipient.

DID also requires the user to maintain line voltage. In the familiar telephone services, the CO maintains the 48 volts necessary for the line to operate, with battery back-ups in case of a blackout. With DID, since the user provides the power source, if line voltage is lost, the telephone company shuts down the line and callers get a busy signal until it is reactivated by the telephone company. At this point in the planning stage, the advice of the local telephone company technician on how to prevent your DID installation from going down will prove useful.

OCR Routing

When the problem of handwriting recognition is solved, optical character recognition (OCR) technology undoubtedly will play a major part in routing. Meanwhile, OCR routing depends on conventions such as code numbers appearing within a specific area of the cover page, combinations of forward and backward slashes (///\\), or bar codes similar to those in use by grocery stores. These options require senders to be trained or use specific formats, at least for the cover page. There are OCR software utilities that can scan any kind of cover page and attempt to find a name, for instance, by attempting to recognize what is written after TO: by mapping it through a stored name pattern. Accuracy is about 80 to 90 percent for a typed document; if unsuccessful, the system defaults to some form of manual routing.

Chapter 9

Application Programming Interfaces (APIs)

Introduction

An Application Programming Interface (API) is software that an application program uses to request and carry out lower-level services performed by the computer's or a telephone system's operating system. For example, for Windows, the API also helps applications manage windows, menus, icons, and other graphical user interface (GUI) elements. In short, an API is a "hook" into software. An API is a set of standard software interrupts, calls and data formats that application programs use to initiate contact with network services, mainframe communications programs, telephone equipment of program-to-program communications. For further example, Standardization of APIs at various layers of a communications protocol stack provides a uniform way to write communications. NetBIOS is an early example of a network API. Applications use APIs to call services that transport data across a network.

There are thousands of APIs. The most relevent to Computer-Based Fax Processing are described briefly.

Telephony Applications Programming Interface (TAPI)

We can easily imagine an office in the not too distant future where different messaging media: voice, text, graphics images and video are integrated and controlled from the desktop PC. To make a phone call, you would pull up a phone book on the screen, highlight the person or company you want to call, and click on the "Call" button with your mouse. To make a conference call with two of your colleagues, you would pull up the company directory and drag the parties' names into the "Conference" box. To check for messages, you would open your on-screen "in-box" to see a list of new and saved voice, text, image and video messages. To reply, you would compose a document in any application, and drag the document to the "Reply" button. The PC might determine the medium and means of communication required to deliver the reply.

Devices such as PBXs, fax machines, desktop telephones and printers have traditionally been built to accept and/or transmit information on their specialized medium, ie. a telephone line or a sheet of paper, but not to provide intelligent, two-way communication with a controlling desktop computer or network. A printer, for example, can only transmit a few rudimentary messages back to a computer: "out-of-paper", or "on/off-line". This lack of communication is despite the presence in most of these devices of sophisticated microprocessors comparable in power at least with early PCs.

In the future, printers will have greatly enhanced abilities to communicate: from your PC, you will be able to see the number of sheets of paper in the feeder, check the level of toner in each color cartridge, send a command to switch to stationery instead of blank paper, and so on. The desktop telephone will have an RS-

232 serial port or other connection to a computer which will enable two-way communication: the computer will be able to ask the telephone to dial a number, for example, and the telephone will send status indications back to the PC similar to those displayed on the phone through colored lights or an alphanumeric LED panel. Ultimately, the telephone itself may become an expansion card with a socket for a handset or headset with microphone and loudspeaker.

The company phone system will be controlled, probably via a serial link, by a PC attached to the corporate LAN instead of relying entirely on its internal CPU. When an incoming call is signaled, the PBX will notify the computer, which will reply with a command telling the PBX where to route the call. (Such PBXs, so-called "dumb switches," are already available from companies such as Summa Four). To make a conference call, the controlling computer will send a command to the PBX. The voice mail PC will be connected to the PBX as a series of extensions and also to the controlling PC via a LAN.

Desktop Applications

Microsoft has a clear vision of how this technology should be tied together: via graphical applications running in a Windows environment. As a first step, Microsoft, in conjunction with Intel, has drafted version 1.0 of the Windows Telephony Applications Programming Interface, or "TAPI", which was published in May 1993. This is an attempt to define a set of function calls (the API) which will be used by Windows programmers to interact with telephony devices such as telephones, telephone lines and telephone switches. TAPI is sometimes referred to simply as "Windows Telephony."

Some important categories of application which will incorporated TAPI include:

Integrated Messaging

A single system for receiving, archiving and responding to voice, text, graphics and video messages.

Personal Information Managers

These will include facilities for automated dialing and collaborative computing over telephone lines.

Advanced Call Managers

Do you really know how to use the conferencing feature on your company telephone system? How to ask for "camp on", where a call will be put through to your colleague who's currently on the phone as soon as he hangs up? How to put that "At lunch" message on the phone display panel to notify callers to your extension that you're away from your desk? I don't know how to use half the features on our digital phone system, and I know I'm not alone. I'm looking forward to an application that gives me access to the features of our phone system through graphical menus on my PC. Through advanced call managers, Microsoft is envisaging control of your phone system through Windows.

Communicating Applications Specification (CAS)

CAS is a high-level API developed by Intel and DCA that was introduced in 1988 to define a standard software API for fax modems. CAS enables software developers to integrate fax capability and other communications functions into their applications.

An important advantage of the CAS interface is the ability to combine ASCII text files with graphics files in PCX and DCX format in creating fax documents. The CAS manager will also automatically construct transmission cover pages, including date, time, sender and message text fields. Another useful feature is the ability to perform a high-speed, error-correcting file transfer to another device supporting CAS.

The CAS programming model deals with events. An event is a single phone call involving the fax board and a remote device such as a fax machine or another fax board. An event is one of the following types:

Send	The computer makes a call and initiates a transmission of one or more files to a remote device (fax machine or fax board).
Receive	A remote device makes a call and initiates a transmission of 1 or more files to the computer. A program can obtain information about receive events which have taken place by querying the Receive queue.
Polled Send	The computer waits for a remote device to call and then starts sending information to it.
Polled Receive	The computer makes a call to a remote device, then receives a transmission from it.

Each event is assigned an "event handle" by the CAS manager for the given board. The event handle will be a number in the range 1 .. 32767, which uniquely identifies the event. No two events for the same board will have the same handle.

Each event has an associated "Event Control File" which contains information about the event, such as the date, time, phone number, file name(s) etc. For events initiated by the computer (Send and Polled Receive), Event Control Files are created by the functions making the request. For events initiated by the remote device (Receive), the Event Control File is created by the CAS manager.

A File Transfer Record is included in the Event Control File for each file transfer operation associated with the event.

An Event Control File created by a Receive event will contain one File Transfer Record for each file received, the CAS manager is responsible for this process.

There are three queues of events maintained as linked lists of Event Control Files by the CAS manager for each installed board:

Task Queue The Task Queue contains an Event Control File for each event which has been scheduled, but which has not yet been executed.

Receive Queue The Receive Queue contains an Event Control File for each completed Receive or Polled Receive event.

Log Queue The Log Queue contains an Event Control File for each event which has been successfully or unsuccessfully completed.

Queues are maintained by the CAS manager sorted in chronological (date and time) order.

An event retains the same event handle even when it is moved from one queue to another. For example, a Send event will be moved from the Task Queue to the Log Queue when it completes, but will still be identified by the same event handle.

A completed Polled Receive or Receive event will appear in both the Receive Queue and in the Log Queue. Two copies of the Event Control File exist in this case — if the event is deleted from one queue, it will still appear in the other.

Other events will appear in one queue only:

Completed Receive and Polled Receive events: Receive and Log Queues.

Pending Send and Polled Receive events: Task Queue.

Completed Send events: Log Queue.

If an event is in the process of being executed, it is known as the Current Event, the Event Control File for the Current Event is not in any queue.

A summary of the CAS functions is as follows:

```
00  Get Installed State
01  Submit Task
02  Abort Current Event
05  Find First
06  Find Next
07  Open File
08  Delete File
09  Delete All
0A  Get Event Date
0B  Set Event Date
0C  Get Event Time
0D  Set Event Time
0E  Get Manager Info
0F  Auto Receive State
10  Current Event Status
11  Get Queue Status
12  Get Hardware Status
13  Run Diagnostics
14  Move Received File
15  Submit Single File
16  Uninstall
17  Set Cover Page Status
```

Parts of this chapter have been adapted from Bob Edgar's book: **PC-Based Voice Processing**, available from Flatiron Press, 800-LIBRARY. No telephony library is complete, without it. I thank Bob, founder of Parity Software Development Corporation for his help and advice with this chapter.

Chapter 10

Fax Modems

Introduction

There is much confusion regarding what the perplexing varieties of Class 1 and Class 2 data/fax modems can and cannot do. There are claims and counterclaims as to whether they are capable of working in a busy fax server environment, where mission-critical faxing takes place.

Anyone considering the use of Class 1 or Class 2 modems for mission-critical faxing must consider the pros and cons of these technologies before making a final decision that may affect an important aspect of the running of a corporations business.

Class 1 Modems

Class 1, which was approved by the ITU-T in 1990, was one of the first specifications for fax communication. It is a series of basic Hayes AT commands used by software to control the board. The first layer, which is addressed by Class 1 modems, operates close to the data link level. At this level, only very simple operations are performed: HDLCs (High-Level Data-Link Control standard.)

As a fax handshake signal format, HDLC frames are used for the binary coded handshaking. They may have a digital identification signal (DIS) describing the machines features, a non-standard facilities (NSF) frame describing vendor-specific features, and a called subscriber identification (CSI) frame containing the answerers telephone number.

A Class 1 and 2 modems contact with the world is through an RS-232 connection serial port operating at a rate of 19.2 kbps or less. This means that every time a byte of data is sent, it must go to the serial port and there are delays inherent with that sending 8 bits at 19.2kbps takes time. It takes even more time to send an HDLC frame or to do the T.30 protocol. For example, when the fax machine at the other ends sends it's DIS, it is received by the modem, which then transmits it over the serial port to the computer, the computer then interrupts whatever it is doing, interprets it, then sends the appropriate command through the port back to the modem. It is important to understand here that the T.30 protocol does not reside in the modem, but in the computer, and that its processing requires the computers undivided attention if it is to direct the modem at each step of the fax communication process.

The Standard That Never Was

When, in 1991, Class 2 was published as a ballot standard by the ITU-T committee, the ballot failed due to a number of technical and political issues. There was much disagreement over what members and industry believed the specification needed to make it work right. The committees debated many issues for over a year before agreeing on what would be in the final Class 2 specification.

Unwilling to wait for the ponderous ITU-T decision-making process to grind on, a number of manufacturers anxious to get into production, took the ballot as if it were a specification and went into production. The result was that after a time there were sufficient Class 2 modems out in the world to make the ballot a de facto standard, although several key areas are undefined and can vary from vendor to vendor.

After the ITU-T solved the issues, they published the final Class 2 specifications, which differed significantly from the Class 2 ballot. In order to distinguish the new real Class 2 specifications from the earlier ballot ones, the ITU-T used Class 2.0 as the new name. Thus there is a fundamental difference between Class 2 modems based on the ballot standard and Class 2.0 modems based on the official ITU-T standard.

One of the most important differences deals with the packet layer protocol in the Class 2.0 specification, guaranteeing more secure throughput over the RS-232 line. Class 2 does not feature this.

Programmers began creating software for the Class 2 modems precisely because they did (and do) exist in great numbers. However, when they want their software packages to be retroactively compatible, programmers discover there is no source document that they can work with, except perhaps for a photocopy of what used to be the interim ballot and nothing else. This is because Class 2 was never approved by the ITU-T. In fact, it is not even possible to get back copies of the original ballot, because once the committee approved the final Class 2.0 specification, all previous work was discarded.

So Class 2 is an illegitimate offspring nobody wishes to recognize and with the Class 2 installed base still out on the world, these questions and problems will not soon disappear.

In the committees view, then, Class 2.0 is Class 2 done right, and it has the ITU-Ts blessing and official industry approval as a standard; that is, there is an official specification available. Unfortunately, since it took the ITU-T over a year to come up with this standard, by the time it was published it was already obsolete in terms of new features added to T.30, which ought to have been included. Not all of the latest fields such as subaddressing and binary file transfer protocols are present in the official Class 2.0 specification.

A revision to the Class 2.0 specification is currently under discussion by the appropriate committees, to bring it up to date with T.30 developments. However, when it will be passed, and what will its final form be, is anybody's guess.

On the positive side of this equation, the next revision, Class 2.01 or 2.1, requires only fairly straightforward changes, such as putting in commands to support the latest features of T.30. This activity is more in the line of mechanical editing than design, and should not excite controversy. It comes down to deciding which command will support a new T.30 option and selecting the right letter for it.

The problem, however, remains that the installed base of Class 2.0 modems is still relatively small, especially when compared to the enormous installed base of Class 2 devices that are already out there.

Class 2 Modems, T.30, and Compatibility

A Class 2 (and 2.0) modem has the T.30 protocol on board. It will do such things as SET DIS and SET CSI. The user can command it to connect to the other station, and it will dial the telephone number, exchange the DIS, the DCS, and preform all of the handshaking operations before sending the data. There is no need on the hosts part to send HDLCs or image, just data. The Class 2 modem will perform the fill bit operation and other T.30 functions that may be needed.

This has a plus and a minus. The plus is that the modem does the T.30 handshaking, with time-sensitive and time-critical functions being performed by it instead of the computer. The minus is that the T.30 protocol is locked in the hardware, therefore any changes to the fax protocol cannot be done simply through software, requiring instead new hardware. In a server environment, image-processing such as ASCII-to-fax, headers, PCX, etc., must be handled by the host.

Another problem is that many Class 2 fax modems are incompatible with one another and other fax devices because there is little cooperation among fax modem manufacturers. This is a key point and must be given careful consideration.

The T.30 protocol has some areas the duration of timeouts, for example that are not specifically defined and are therefore subject to interpretation.

During the 1984 to 1986 time frame, there were very few T.30 fax protocol engine engineers in Japan, and none in the U.S. Although they competed on manufacturing, performance, and price, all Japanese fax manufacturers cooperated with each other to ensure that all manufacturers could send and receive faxes to and from other machines. Precisely because it is open to interpretation, this rendition of the T.30 protocol was not tampered with, ensuring compatibility. Since there was only a handful of manufacturers, this was not too difficult an enterprise. Now, however, it is an entirely different story.

At present, the majority of those involved in the manufacture of fax modems have produced fax protocol versions based on their individual interpretations of the T.30 recommendation. Although this may occasionally result in what is boasted of as a superior fax protocol, the problem still remains that it will not talk to all the fax devices machines and boards out in the world.

There is a significant number of incompatibilities for Class 1 and Class 2 modems and the general population of fax devices in the marketplace. End-users and even developers have very little recourse, unless they actually write T.30 code themselves, but then they are burdened with that task of fixing (and what may need fixing is not always obvious) something that does not work at all or only partially. In the end it may still be incompatible, only now with an entirely different set of fax devices. The level of T.30 compatibility testing that the fax board and machine companies carry out is unknown in the Class 2 fax modem environment.

The problem has been exacerbated by increasing competition. Clearly, from a compatibility point of view, two fax modems from the same manufacturer can probably talk to each other, but there still may be other problems such as host computer dependencies and timing issues. (For example, a modem may not operate exactly the same in different computers.) Timing issues are critical because, as asynchronous device, a modem expects data when it requests it. If the host computer is not immediately there to provide it and a piece of data is missed, a scanline or error correction mode (ECM) train dropped, the damage is done and the fax does not come out straight or, if ECM is present, it becomes necessary to retransmit.

ECM is available in Class 2.0, but not in Class 2. This is one of the reasons why many developers working with Class 1/Class 2 modems generally drive them in Class 1 format for control (Class 2 supports Class 1 operations), and do all T.30 operations on the host computer.

With Class 1 almost everything is done by the host computer, where as with Class 2 some of it is off loaded to the modem, such as some of the buffering and timing responsibilities.

Multiple-Modem Applications

A Class 2 modem has the capability to buffer a few seconds of data. So if some data is missed, it just slows down a bit while making it back up. Because of this, a Class 2 is better able to handle a busy host, and it would be theoretically possible to run multiple Class 2 modems on a single host and a fax server. It is extremely unlikely to run multiple Class 1 modems because of the interrupts the need to service the data is almost constant.

With a Class 2 modem the problem of needing to adjust to the capabilities of the remote device is not solved. The information about various compressions (MH, MR, MMR) is still on the host and carried out by it, and it has to provide the right data to the modem. So steps such as conversions from ASCII into fax are still being performed by the host, through a serial port.

Serial Ports

A DOS serial port is interrupt-driven (and therefore slow for fax purposes,) with each interrupt occurring on a character-by-character basis. To properly keep up with fax it is necessary to run at a greater rate: one equal to the faxes. If the rate equals the fax speed, eventually this results in underrun situations. Another point to consider is that, as an asynchronous interface, a DOS serial port uses ten bits per character instead of eight. These additional two bits, a start and stop bit, can consume considerable CPU time. A fax transmission running at 9600 bits per second requires just that: 9600 bits every second. However, if a DOS serial port is used, that means 9600 plus a 25 percent timing overhead to accommodate start and stop bits.

Thus, the only reasonable solution to using a 9600 bps fax port this way requires that the serial port be run at 19.2 kbps. That requires an interrupt every 400 to 500 msec, which means the CPU is doing an interrupt of service routine with programs every 400 to 500 msec to read in the next character to the serial port. Therefore, any CPU activities that are time-critical, such as switching memory banks, going in and out of 386 modes, etc., which Windows is very prone to do, can severely impact the faxing as well as the program processing capabilities of the modem and computer. Also, any other operations that are interrupt-free, such as screen rights or display updating, may cause the serial port to experience severe character losses. The interrupt overload also degrades a PCs general performance by forcing it away from other tasks to service an interrupt and then returning to continue the previous task. This kind of back and forth and back and forth switching carries with it considerable overhead.

While some of the more modern UARTS pipelines to compensates lightly for this problem, providing a little more tolerance towards these delays, the operation is still very time-critical. In the case of Windows 3.1, the program simply does not work well at these rates of speed. Anyone implementing an RS-232 driver running at 19.2 kbps under Windows knows this implementation will be very difficult and unreliable. In fact, Microsoft is in the process of rewriting its asynchronous drivers for future versions of Windows to get around the problems that take place under version 3.1.

Conclusions

What an end-user needs, must be determined by use; how many faxes a day, a week, a month are sent and received. For the desktop, laptop, or casual user there is nothing wrong with Class 1 or Class 2 modems. There may be compatibility issues but there is usually a fax machine handy if a fax has a problem in getting through. Most of the Class 1 and Class 2 fax modems sold today have never sent or received a fax. They were bought for data. In fact, nowadays, anyone looking for a data modem is hard-pressed to find one that does not offer a fax capability of some sort. Computer store salesmen are fond of saying, *"buy a modem and get fax for free."* This, however, is one time when it may be useful to look that gift horse in the mouth.

System integrators and ISVs build mission-critical fax applications. If some of their customers send hundreds or thousands of faxes a day, it is crucial that the faxes get through, and that the user know if a specific fax did not, and if not, why. These users cannot afford to deal with incompatibility tie-ups. For vital applications of this kind, the price of the fax board, whether $139 or $1,000 becomes a much less significant variable in the systems cost. That is the market high end fax board address: one in which getting the fax through is critical, where keeping track of expenses is critical, and where sending faxes at the lowest transmission costs as in the case of high-volume applications is an important goal. In terms of the occasional fax sent from the laptop or standalone PC, the user maybe better off by buying a data modem and getting fax for free, always remembering, however, that one gets what one pays for.

Chapter 11

Computer Telephony Busses

Introduction

What are computer telephony buses? A bus is a familiar term in the computer world, and means a electrical, mechanical, and signalling protocol to allow different components within a computer to work together. Examples of buses in the computer world include ISA bus, EISA bus, sbus, VMEbus, etc.

The computer telephony industry has adopted the use of buses for the same reason - to allow different computer telephony components to work together in the same system. Unlike the "host bus" that allows these expansion boards to communicate with the computer's CPU, a separate bus for carrying real time information between various computer telephony devices is critical to making these devices perform as a complete system. In a standard IBM PC environment, these buses are physically implemented as a mezzanine bus, and the physical connect consists of one or multi-drop or point-to-point ribbon cables connecting the computers together. On other platforms, the physical transport could be the same at the host bus, like the VME platform or a proprietary switching platform.

This chapter will explore the four different kinds of buses that have arisen in the computer telephony world - the Analog Expansion Bus (AEB) (tm), the PCM Expansion Bus (PEB) (tm), the Multiple Vendors Interpretation of Protocol (MVIP) (tm) bus, and the Signal Computing Bus (SCbus) (tm).

Common Elements

Despite the different architectures and protocols of the four standard buses, all were designed to accomplish the same purpose. Each computer telephony bus allows the system integrator to extends the capability of a single computer telephony component by allowing multiple components to work together in the same telephone call. There are two basic reasons for this:

The first reason to use a ct bus is when the telephone network interface is physically distinct from the signal computing component (voice processing, fax, speech recognition, ...). As discussed in chapter xx, there are large variety of ways for a computer telephony system to connect to the public switched telephone network, including analog loop start or direct-inward-dial, and digital T-1, E-1, basic rate ISDN, primary rate ISDN, or proprietary PBX links. By separating the network interface from the signal computing component, system integrators can mix and match different components to build the system they need. In addition, different vendors can supply different components of the total system.

The second reason to use a ct bus is when the system integrator wants to augment the capabilities of computer telephony system. For example, some installations might call for speech recognition capabilities; others may not. By employing a ct bus, both types of systems can use the same basic architecture and most of the same components. That system that requires speech recognition simply augments its system configuration by adding a speech recognition resource via the computer telephony bus.

In addition, the new generations of buses, like the SCbus, have a large inherent switching capability which greatly increases the switching fabric within a computer telephony system, and in many cases can be used to replace an external switching device.

Analog Expansion Bus

History

The AEB is the first open computer telephony bus, and was introduced by Dialogic Corporation on their D/xx line of voice processing products. In order to spur industry growth, Dialogic publishes the specifications to the AEB, and provides technical document for the bus at a nominal charge. The AEB was designed to allow system integrators to attach Expansion Modules, like a facsimile component, that extend the basic capabilities of the D/xx, and to allow system integrators to attach digital Network Interface Modules for use with the D/xx.

The AEB was eventually adapted by the other leading voice processing manufacturers - Rhetorex and Natural Microsystems and by fax vendors GammaLink and Brooktrout.

Functionality: The AEB supports four voice processing channels via a 20 pin connector. Each of the four channels has a audio signal, and signaling transmit and receive signals.

101

The audio signals are terminated, 2-wire, bi-directional access points. The signaling states are carried on TTL logic level signals to allow software-transparent (to the application program) signalling connections between different computer telephony modules connected to the AEB. On/off hook commands and network alerting (ringing) or off-hook complete (using T-1 E&M protocol) signals are passed via the signaling pins.

Advantages and Disadvantages

The AEB is simple, inexpensive computer telephony bus that has been adopted by the leading computer telephony for low density systems.

It is not well suited, however, for high density systems. The analog nature of the bus does not lend itself to the multi-drop or high density nature of many larger computer telephony systems, which is why Dialogic introduced the PCM Expansion Bus.

Sample of products that support the AEB

Voice processing: Dialogic D/21D, D/41D, D/41E
                      Natural Microsystems
                      Rhetorex

Fax: Brooktrout TR-112
                      GammaLink CP4/AEB
                      Rhetorex RFAX/4000

Speech Recognition: Dialogic VR/40

The two most common configuration examples are: a) Voice processing card to resource module (fax or speech recognition) and b) Digital network interface to resource modules.

PCM Expansion Bus (PEB)

History

The PEB was introduced by Dialogic Corporation in 1989. It is the first digital, open computer telephony bus and it helped the industry expand into higher density systems. The PEB used time division multiplex techniques to pass digital, pulse code modulated date (hence its name) between computer telephony computers.

In order to spur industry growth, since 1991 Dialogic has published the interface specifications to the PEB and does not require a license for its use. Detailed technical documentation to the PEB is available from Dialogic at a nominal fee. Given Dialogic's market position, a large number of computer telephony component providers have PEB based components.

Functionality

The PEB is a 24 pin, high speed, digital, TDM bus that, depending on the clocking rate, can support 24 or 32 simultaneous telephone channels. To support T-1 and 24B+D PRI applications, the PEB is clocked at 1.544 Mbps and supports 24 full duplex channels. To support E-1 and 30B+D PRI applications, the PEB is clocked at 2.048 Mbps and supports 32 full duplex channels.

The PEB world is divided into two types of components - Network Interfaces attach to the telephone network, and typically control the clocking of the PEB, and Resource Modules that perform some signal computing function (e.g. voice processing). Resource Modules switch onto and off of the PEB via time slot assignment - each resource module is capable of transmitting on and/or receiving from some range of time slots on the PEB.

Advantages and Disadvantages

The PEB is a widely embraced bus. Many companies have PEB-based product on the market, allowing developers to mix and match pieces to serve a broad range of applications. The PEB is relatively easy to implement, and takes little material cost. Dialogic offers the technical specifications for the PEB at virtually no cost, and offers design assistance.

The PEB, though, is a first generation bus. It relative lack of time slots is a disadvantage in high density systems.

Sample products that support the PEB:

Voice processing: Dialogic D/121, D/240-SC, D/300-SC

FAX: Brooktrout TR118
                       GammaLink CP4-SCOAZ xxx

Voice Recognition: Dialogic VR/160p
                       Dialogic VR/120VPC VPRO4

Switching: Amtelco
                       Dianatel xxx
                       Dialogic MSI/C and DMX

Network Interfaces: Acculab xxx
                       Dianatel xxx
                       Dialogic DID/120, LSI/120, LSI/80.
                       Promptus xx

Text-to-Speech: Dialogic TTS/120

Configuration Examples

The most common configuration example is that of a network interface connected to one more resource modules. Since each resource module can select transmit and receive time slots, different resource modules can participate in the same call, even at the same time. For example, a T-1 network interface (e.g. DTI/211) might be connected to 24 channels of voice processing (e.g. D/240-SC), four channels of fax (e.g. CP4-SC), and 12 channels of speech recognition (e.g. VR/120p). Depending on the needs of the caller of the system, the appropriate resource can dynamically connect and disconnect on the call.

Another popular configuration is the drop-and-insert configuration. In this configuration, the call processing unit fronts end a switch. The front-end call processing unit intercepts some or all of the calls destined for the switch, and provides call processing resources for the call. The call processing unit might, for example, query the caller for an account number or offer to automatically fax information to the caller. If the caller needs to reach a live operator, the call processing unit can pass the caller through to the back end switch, and on to an operator. Any information the call processing unit collected from the caller, like an account number, can be passed to the operator or an associated computing device.

To implement a drop-and-insert configuration, a network interface is placed on both sides of the PEB cable, both logically and physically. In between, there is a cross-over cable. Automated call processing equipment can be placed on one or both sides of the cross-over cable.

MVIP

History

The Multi-Vendor integration Protocol was announced in 1990 by Natural MicroSystems along with Mitel Semiconductor, GammaLink, Brooktrout Technologies, Voice Processing Corporation, Scott Instruments and Promptus Communications.

MVIP was presented as an open standard for interoperability among telephone-based resources, including trunk interfaces, voice, video, fax, text-to-speech, and speech recognition, to enable design of larger,. multi-technology computer telephony systems.

Functionality

MVIP defines intra-node and inter-node digital communication buses and defines multiple switching models for allowing devices to communicate in real time.

Intra-Node Bus

The MVIP Bus specification is based on Mitel's ST-Bus, and is a digital telephony bus that carries PCM data between devices. The MVIP Bus consists of 16 serial data lines clocked at 2.048mhz, providing 512 half-duplex timeslots, or 256 full duplex timeslots. Since transmit and receive timeslots are allocated in pairs to each resource (channel on a device), the MVIP bus can support nonblocked communication among up to 256 resources.

The bus can be implemented on ISA, EISA, or MCA platforms as well as other computing platforms. In a PC implementation, the MVIP bus would be implemented using a ribbon cable with a 40 pin connector. If it is necessary to transfer signalling information between devices, the MVIP recommendation is to carry this information through associative channel signalling protocol, using data timeslots. In this case, the number of available timeslots for data transfer is further reduced to 128.

Multi-Node Systems

MVIP defines a specification for interconnecting individual nodes into larger, distributed systems. Called Multi-Chassis

MVIP, this specification was presented in 1993 and currently proposes 4 different implementations. These are as follows:MC-1: up to 1408 timeslots running at 4.096 Mbps over twisted-pair cableMC-2: up to 1536 timeslots over FDDI-II or copperMC-3: up to 2300/4600 timeslots over SDH/SONET fiberMC-4: (ATM) no details published

Detailed specifications are not publicly available and are not included in the MVIP Reverence Manual.

Switching

The MVIP architecture assumes that most traffic in most MVIP applications is being connected between network interfaces and call processing resources. Devices are differentiated by whether they have telephony interfaces or whether they are doing processing only (ie, voice store and forward, fax, or ASR). In the 3 switching models originally specified in MVIP, a telephony interface device or a switching device was required to have some level of switching capability and that this device would switch information between the various resources involved in processing a call.

The level of switching affects the maximum system size and configuration. The three levels are:

MVIP Switch Compatible

The lowest level of switching requires that the device maintain a switch capable of making full-duplex connections of any incoming telephony channel to a subset of MVIP bus timeslots and vice versa. In this case, a device would only be capable of switching information among a limited set of resources.

MVIP Standard Switching Compliant

This level of switching allows the device to switch data from the network interface to any MVIP bus input or output timeslot consuming only 4 timeslots on the MVIP bus (2 transmit and receive pairs). In order to switch information to another channel on the same device, an MVIP Standard Switch Compliant device would use 2 MVIP bus timeslots.

MVIP Enhanced Compliant Switching

An Enhanced Compliant Switching device is capable of switching data from the network interface to any MVIP input or output timeslot. Because it has additional switch paths to connect incoming and outgoing network channels within a trunk interface, an Enhanced Compliant device can connect to any other channel on the same device without consuming any MVIP bus timeslots. An enhanced switch compliant device could serve as a central switch for an entire MVIP chassis.

With the new FMIC (Flexible MVIP Integrated Circuit) chip, it is possible to have switching completely distributed so that information could be switched directly between resources. However, this is not recommended in the specification and at present many devices do not yet use the FMIC.

Advantages and Disadvantages

Advantages: Higher capacity: The MVIP Bus offers higher capacity than the PEB which requires a separate switching device for more than 24 or 32 channels.

Choice of Implementations

Telephony interface manufacturers can support different levels of switching compliance depending on system size and configuration. This allows them to economize on cost depending on the intended application.z Products: MVIP claims approximately 100 hardware products are available.

Disadvantages

Timeslot allocation: Because MVIP requires fixed transmit and receive timeslots for each processing resource, certain applications like broadcasting data to multiple resources also requires multiple timeslots to duplicate the data for each listener. Depending on the level of switching compliance, additional timeslots are required to route calls between telephony interfaces for configurations like drop and insert.

In-band signalling: Signalling information or messages between MVIP devices must be carried on the host bus or in-band using data timeslots. If this information is carried in band, this halves the effective system capacity.z Lack of standard APIs: Because MVIP does not define a standard software architecture, it is difficult to design complex systems using products from multiple vendors, each of which use proprietary APIs.

Sample products that support the MVIP (Note: this is not a complete list. For a comprehensive list of actual hardware products and the level of switching compliance, contact Natural MicroSystems.)

Voice processing   Linkon
Natural Microsystems
Pika
Rhetorex
Bicom

FAX: Brooktrout
GammaLink
OAZ Communications

Speech Recognition: Scott Instruments
Voice Processing Corporation

Switching and Conferencing:

> Amtelco
> Dianatel
> Excel
> MultiLink

Station Interfaces: Voice Technologies Group

Network Interfaces: Acculab
> Dianatel
> Telesoft
> DataKinetics
> Link Technology
> Promptus Communications
> SCii

THE SCSA Hardware Model

History

The Signal Computing System Architecture (SCSA) is a hardware and software architecture launched in March 1993 by Dialogic Corporation and 70 other telecommunications and computing equipment suppliers including Digital Equipment Corp., Northern Telecom, NEC, and IBM to simplify the task of building larger, more complex computer telephony systems using multiple technologies.. SCSA active supporters currently number 240+ companies, many of whom are participating in defining the SCSA specifications.

The SCSA Hardware Model defines the hardware portion of the SCSA architecture. It consists of a real-time digital communications bus with a separate messaging channel, a multi-node network architecture for development of multi-node systems, and a single, distributed switching model. Servers using the SCSA Hardware Model can be controlled using various programming models including the SCSA Telephony Application Press RETURN for more; type NO to stop:

Objects (TAO) Framework or through other vendor -specific software models.

Intranode Bus

The SCbus is TDM bus for computer telephony consisting of 16 synchronous serial data lines for real time communication among devices in a single node and a dedicated messaging channel for carrying signalling and messages between devices. The SCbus supports 1024 bidirectional 64 Kbps timeslots in a mezzanine bus implementation running on a ribbon cable, or 2048 time slots when running in a backplane implementation. An SCSA hardware implementation specification for the VMEbus has been endorsed by VITA (the VMEbus industry trade association) and is currently in submission to the American National Standards Institute (ANSI). The SCbus can also be adapted to other platforms, for example, as part of the backplane of a proprietary switch.

The optional messaging channel provides a means for real-time communication of messages and signalling information among resources "out of band". The diagram shows the message channel implemented as a dedicated bus controlled by a high-level data link controller (HDLC) chip. The message bus allows devices to communicate this information directly without going through the host processor or application for faster system response time. Additionally, developers can integrate specialized resources that use the message bus to get command control and status information. Communication over the message bus is faster and more efficient than embedding the message handling in the data stream and does not consume data timeslots of the SCbus.

The message channel can be implemented using any transport

capable of transferring data in real time. . For example, a developer could implement the message channel over an Asynchronous Transfer Mode (ATM) network where certain packets would be dedicated to messaging.

Multinode Expansion

The Multinode Network Architecture (MNA) provides the ability to connect multiple nodes to build large systems. The MNA is application independent; its connections are transparent to the application. The MNA provides the capability to incorporate nonSCSA nodes. For example, it can be used in high-density PBX integrations to provide a gateway function that extends the SCbus to other nonSCSA products.The MNA can be implemented through various media, depending on the system size requirements and physical location of nodes. Below are 2 examples:

SCxbus

Connects up to 16 co-located nodes using a ribbon cable, for systems up to 1344 ports. This implementation is physically compatible with MC-1.z ATM: An ATM switch can be used to connect larger numbers of geographically distributed nodes for multiple thousand line systems. In this case, the messaging channel can be implemented through dedicating ATM packets for this information.

SCSA Switching Model

SCSA defines a single, completely distributed switching model in which all devices are peers. It features flexible time slot allocation where all devices can transmit or receive on any time slot on the bus, enabling communication between any devices in the system. Other features include automatic clock fallback and switchover for fault tolerant application requirements, time slot bundling with full frame buffering for technologies like video, and broadcast capability where one device can transmit to multiple devices while consuming only 2 timeslots regardless of the number of "listeners". This model is easily scalable and suitable to implementation on different hardware platforms, for example, from a PC ISA bus to VMEbus.

Switching is controlled through the SC2000 chip, designed by Dialogic and built by VLSI Technology. The SC2000 also offers compatibility modes for other buses including PEB, MVIP, and ST-BUS, allowing developers to design hardware that with different software loads can operate in SCSA, PEB, or MVIP system.

Advantages and Disadvantages

Advantages: Higher capacity: The SCSA Hardware Model today offers the highest capacity and fastest speed of any computer telephony architecture, for the greatest degree of scalability.

Compatibility modes: Because the SC2000 chip offers compatibility modes, developers building hardware devices for multiple hardware platforms can economize by building a single hardware product that can run in different modes, with the appropriate software.

Single switching model: Because all SCSA products implement the same switching model, there are no limits on device location, enhancing system scalability. It also provides the most efficient allocation of timeslots and eliminates the need for dedicated switching hardware.

Message channel: The separate message channel provides faster system response time especially in client server configurations where applications may be running on a remote platform. The message channel also supports applications where out-of-band signalling capability is required without consuming any data

113

time slots.

Defined software architecture: Because SCSA also defines a comprehensive software architecture (SCSA TAO Framework), the hardware architecture is designed with consideration for application development concerns like performance features required for client server environments. SCSA TAO Framework offers several capabilities critical to supporting distributed computer telephony systems like the ability to support multiple applications, dynamic resource allocation, and standardized vendor-independent application programming interfaces.

Disadvantages: As a newer architecture, components that support the SCSA Hardware model are still being released. Several companies are still designing products which will be released in the upcoming months.

Sample products that support the SCSA Hardware Model Currently announced products include:

Voice processing: CallTech
                     Dialogic

FAX: GammaLink

Speech Recognition: Voice Control Systems
                     Processing Corporation
                     Telefonica de EspanaDasa
                     VSPhillips Professional Systems

Switching and Conferencing: Amtelco

Station Interfaces: Digivox

Network Interfaces: Acculab
                     DataKinetics
                     Dialogic

Discofone
Telesoft Design

Text-to-Speech: CSELT
Elan Informatique
Lernout & Hauspie

Summary and Recommendations

Computer fax board buyers should give careful consideration to the architectural platform when choosing a product supplier. Although many computer telephony technologies are still in an evolving state, the demand for integrating these "media processing services" with facsimile processing is growing, because there is a fundamental need to make "bridges" between once separate communication tools like telephones, facsimile and workstations in order to bring information to users in the most useful form.

In choosing an architecture, the user should also evaluate the software architecture, which can make a great difference in the time needed to bring their systems to market. The software architecture must be able to support client server considerations like support for multiple applications. Physical connectivity between hardware alone is of limited value if the products cannot provide features like multi-application support and dynamic allocation and sharing of resources.

Chapter 12

Fax Boards
GammaLink Product Line

Introduction

CBF boards, like the ones manufactured by GammaLink, enable users to receive, print and store faxes. Depending on the type of system configuration, these boards can act as fax servers for vast computer networks, sending, receiving, and routing faxes for entire multi-national companies.

GammaLink originated PC fax technology in 1985 with the release of the first PC-to-fax hardware and software products. Since that time, GammaLink remains a leading vendor in the computer-based facsimile (CBF) communications market. Although GammaLink no longer addresses low-end boards, the

117

company offers hardware and software for networks of all sizes and complexities.

An increasing number of developers now use GammaLink boards to create systems, adding them to their proprietary hardware and software products.

Originally, fax boards were viewed simply as a means for stand-alone PCs to emulate fax machines, and applications were limited to the transmission of ASCII files. Although fax transmission has been always used for highly visual data, poor quality reproduction was unavoidable until the advent of CBF.

GammaLink, and its family of computer-based fax products, provides solutions which meet the diverse requirements for fax integration. Users can select single and multi-line fax boards, and have the choice of working with the Analog Expansion Bus (AEB), Multi-Vendor Integration Protocol (MVIP), or Loop Start Interface (LSI) for voice and LAN-based applications. Most importantly, GammaFax boards provide users with the flexibility and robust reliability required for high volume fax traffic. The following describes GammaLink's most popular fax boards and products:

GammaFax XPi

The GammaFax XPi is a network-ready fax board and communications software. Considered the price/performance leader in the network fax board marketplace, it offers four-line capability, broad software support, and is compatible with any PC-based LAN. Designed for the corporate environment, the XPi allows users to match the company's networking capabilities with its fax requirements. All GammaFax menus and utilities are supported by the XPi.

The GammaFax XPi and CPi (see below), are the first multiline facsimile boards designed with the throughput and robustness required for network use. The 16-bit microprocessor carries out all communication operations, text, and PCX file-conversion while transmitting at 9,600 bps. Up to four GammaFax XPi or 16 CPi boards can be installed in a single PC for maximum throughput on even the largest networks.

Features of this system when used with fax-sever software are easy-to-use pop-up menus, background transmission and reception, broadcasting to distribution lists, micro channel or standard bus, high quality output, and a transparent growth path. Programmability is possible with the GammaFax XPi.

GammaFax CPi

The GammaFax CPi is the top-of-the-line single-port PC fax board capable of supporting up to 16 lines in one chassis, and is targeted at the fax server, fax switch and commercial fax service environments. The CPi has a 16-bit microprocessor with 512K of RAM enabling sixteen boards to be installed in a single chassis, all sending or receiving at 14,400 bits per second simultaneously.

The GammaFax CPi communications software supplied with the board (and all GammaLink hardware offerings) includes advanced management features.

The Queue Manager schedules background fax transmissions, and the communications software provides unique logging and diagnostic capabilities which allow for detailed billing and usage reports.

The GammaFax CPi software also includes the GammaFax Communications Language (GCL.) GCL is a dBase-like programming language that enables VARs and systems integrators to easily integrate fax capabilities into applications. In addition to GCL support, the GammaFax CPi also supports GammaFax Programmers Interface (GPI) for developers to build sophisticated fax switches, develop application software that supports fax capabilities, develop fax servers, and E-mail gateways.

GammaFax CPD

The GammaFax CPD is a network-ready fax board and communications software, for use with Direct Inward Dial circuits. GammaFax CPD is an answer-only version of the GammaFax CPi that supports inbound routing of faxes across a network through the use of (DID).

The GammaFax CPD is the first DID PC fax board to support the ITU-T Error Correction Mode (ECM) for transmitting and receiving faxes with zero errors. It is designed for installations where automatic routing to individuals on a network using DID is desired as well as for custom applications that will use the DID capabilities in a polling application for information distribution to the calling fax machine.

The GammaFax CPD has a 16-bit microprocessor with 512K of RAM enabling sixteen boards to be installed in a single chassis, all simultaneously receiving at 9,600 bits per second. It includes communications software, the GammaFax Command Language, and a power supply.

Because it is an answer-only board, the GammaFax XPi or CPi board is required in order to send faxes from the system.

GammaFax Multi-Line Communications Professional (CP-4/AEB)

The CP-4/AEB is a four-port programmable fax board, designed for voice/fax or high volume T1 switch applications. Each port has a 16-bit microprocessor and 512K of RAM to provide the performance required in high-volume T-1, T-3, ISDN, fax broadcast, interactive voice/fax, and fax-on-demand systems. The CP-4 supports up to eight boards in a single chassis, is programmable, and is compatible with any Analog Expansion Bus (AEB) boards.

GammaLink CP4/LSI

The GammaLink CP4/LSI allows users to take advantage of all the best computer based fax features from one board. The CP4 provides four independent fax channels per board with a loop start telephony interface, one Rockwell modem dedicated to each fax channel, and one 15 MHz microprocessor per fax channel. In addition, the CP4 offers 14.4 Kbps transmission for sending and receiving faxes, providing up to 50 percent speed improvement compared to the 9600 bps products of other competitors. Both v.17 and v.33 modulation are supported on the CP4/LSI, allowing faster transmission speeds over noisy telephone lines which can reduce costs if users frequently send faxes internationally.

The CP4/LSI provides several user-friendly features. For example, the CP4/LSI allows users to minimize their telephone costs with 2D compression (Modified READ) schemes that increase the speed of sending a page by up to 25 percent, and Group 4 compression, which increases the speed of sending a document by up to 40 percent. Users also will benefit from the CP4/LSI's ITU approved ECM (Error Correction Mode) feature which reduces costs by terminating a transmission if the phone line quality is poor rather than send an unreadable fax. In addition, CP4/LSI had the ability to decode and store incoming TouchTone (DTMF) digits, giving users the option of inbound routing using push-button telephone pad commands.

GammaFax CP-4/PEB

The CP-4/PEB board, is an interface card which allows the GammaFax CP-4 to operate over the Dialogic PCM Expansion Bus (PEB). It is used in voice/fax and fax service applications.

GammaPage

GammaPage is GammaLink's PostScript, LaserJet, and CAD translation software. It translates page description languages into fax format, allowing users to fax documents exactly as they have been created. It consists of two software modules: GammaScript allows users to fax documents in the PostScript page description language; GammaJet translates documents in the HP LaserJet (HP-PCL) language. This gives faxes the same high quality expected from a laser printer.

GammaPage is a key technology for making PC-to-fax a widely used medium for telepublishing. The product provides the 35 PostScript typefaces necessary for full compatibility with the Apple LaserWriter NT or an HP Laser Printer. It is supported by all GammaFax boards.

GammaFax Programmers Interface

The GammaFax Programmers Interface (GPI) is a C programmers toolkit for GammaFax CPi and CPD, available for both DOS and OS/2 programmers.

This set of C language subroutines enables OEM developers to incorporate sophisticated fax capabilities directly into applications. For example, an application program that generates updated price lists from a database can interface to a GammaLink product for automatic transmission. The GPI is available for either DOS or OS/2 applications.

Chapter 13

Application Generators and Parity Software's VOS Programming Language

An Application Generator, (also referred to as Ap Gens) is software that writes software. Application generators are software tools that, in response to your input, write software code a computer can understand. Application generator shave three major benefits: 1) They save time. You can write software faster. 2) They are perfect for quickly demonstrating an application. 3) They can often be used by non-programmers.

App gens have two disadvantages: 1) The code they produce is often not as efficient as the code produced by an experienced programmer. 2) They are often limited in what they can produce.

Application generators tend to be general purpose tools. Alternatively, they may be very specific, providing support for specific applications, such as connecting voice response units to mainframe databases, voice messaging system development, audiotex system development, etc. Application generators are often used in programming voice processors.

One of the application generators' bigger advantages is their ability to translate user specified screens and menus into programming code. In essence, you produce the screen or menu using an interface as simple as a word processor. Then the applications generator translates that screen into programming code in a language, such as "C." Once translated into "C", a proficient programmer could go through the code and "improve" on it.

Using VOS To Create Fax Application

Parity Software's (870 Market Street, Suite 800, San Francisco, CA 94102, 415-989-0330) VOS programming language supports the major Dialogic and Dialogic-compatible fax boards using simple function calls. Dialogic-compatible fax boards include those from Gammalink and Intel. VOS is an applications-oriented language which is at a similar level to languages such as dBASE, Clipper and BASIC. VOS includes built-in functions to control the PC screen and keyboard, read and write DOS files, access databases and LANs and to control almost any feature of any Dialogic in any configuration.

VOS simplifies programming greatly compared with the C language for two main reasons. First, VOS is a much simpler language than C and can easily be learned by people without "heavy" programming experience. Second, VOS provides a "multi-tasking" environment where several VOS programs can be executing at the same time, even on an MS-DOS computer. This means that the application developer can write programs to control one caller on one phone line without taking care of activities on many lines at once. In other words, you don't have to use difficult techniques like state-machine programming.

To send a fax using a Gammalink board requires just a few function calls. The following VOS program faxes the C:\AUTOEXEC.BAT file:

```
program
        sc_wait(1);
        sc_offhook(1);
        Gset(1, 2, "C:\AUTOEXEC.BAT");
        Gstart(1, 1);
endprogram
```

As another example, the following function call is all that is required to receive a fax and store it in a TIFF file using a Dialogic FAX/xx board:

```
fx_recv(line, "FAX.TIF", 1);
```

The three arguments to `fx_recv` are:

1. The FAX/xx channel number.

2. The file name to store the fax (may include drive and path components).

3. The file type. This may be 0 for a "raw" fax file, or 1 for a TIFF/F file.

To give an example of a complete VOS program which uses a FAX/xx board (could be a FAX/40 or FAX/120), the following program waits for a call from a fax machine, receives a fax, then re-transmits the fax to a given telephone number. This is a simple example of "fax store-and-forward":

```
dec           # Declare variables, name : length (char)

      var line : 2;
      var filename : 12;
      var code : 3;
      const FAX_PHONE = 5551212;
enddec

program
      line = 1;                 # Hard code line 1 test
      sc_trace(line, 1);        #    Trace     speech     card
events..
      fx_trace(1);              # ..fax events for debug
      filename = "FAX" & line & ".TIF";
      sc_onhook(line);
      sc_wait(line);            # Wait for a call
      sc_offhook(line);         # Answer call
      sleep(10);
      fx_setstate(line, 0);     # Become a Called
      code = fx_recv(line, filename, 1);
      if (code <> 2)
      vid_write("Receive failed");
      restart;
      endif
      vid_write(fx_att(line, 6), "pages received");
      sc_onhook(line);          # Disconnect call
      sleep(30);
      sc_offhook(line);         # Start new call
      sleep(20);                # Wait for dial tone
      sc_call(line, FAX_PHONE);
      if (sc_getcar(line) <> 10)
            vid_write("Call not completed");
            restart;
```

```
      endif
      fx_setstate(line, 1);   # Become a Caller
      fx_file(line, filename, 1, 0, -1, 0, 0, 0, 0);
      code = fx_send(line, 0);
      if (code <> 1)
            vid_write("Send failed");
            restart;
      endif
      vid_write(fx_att(line, 6), " pages sent");
      restart;
endprogram
```

The functions named sc_... are used to control a Dialogic speech card such as a D/41 or D/121 which is used to answer the call, and could in addition be used to play menus to the caller and collect touch-tone responses in order to select a document. Some typical sc_ functions include:

sc_onhook(line)

>   Puts line on-hook ready to receive a new call.

sc_wait(line)

>   Waits for a new in-coming phone call on the given line.

sc_offhook(line)

>   Answers call by going off-hook ("picking up the phone").

sc_play(line, filename)

>   Play a pre-recorded audio file to the caller.

sc_call(line, phone_nr)

Dials a phone number and performs call progress analysis to determine if the call was answered or if a busy tone or other result occurred.

The important fax-related functions are:

fx_setstate

Determines whether the fax board should behave as a Caller (the device which initiated the call), or a Called (the device which received the call).

fx_recv

Analogous to **sc_record**, saves received fax data to a given file name.

fx_file

Opens a given file in preparation for a fax transmission. Several files may be included in one transmission.

fx_send

Analogous to **sc_play**, transmits fax data from one or more files opened by **fx_file**.

The function **sleep** is used to suspend a VOS program for a given length of time, expressed in tenths of a second. The **vid_write** function writes to the PC screen, and is handy for displaying messages.

For another example, the following complete VOS program faxes the AUTOEXEC.BAT file using a CAS-compatible board such as the Intel SatisFAXtion:

```
program
        Fsub(1, 5551212, 0, 0, "",
"C:\AUTOEXEC.BAT");
    endprogram
```

This chapter was provided by Bob Edgar, founder of Parity Software Development Corporation. For more information about VOS, Bob's book: **PC-Based Voice Processing**, is available from Flatiron Press, 800-LIBRARY. I thank him for his help and advice with this chapter.

Section V

**Building a 24 Port PC-Based
Fax Broadcasting System
Using GammaLink Equipment**

Building a 24 Port PC-Based Fax Broadcasting System Using GammaLink Equipment

Introduction

Fax broadcasting is one of the most fascinating and powerful fax technologies available today. It facilitates the delivery of time sensitive data to a user's desktop or laptop instantly. Fax broadcasting in it's simplest form is the transmission of a single document to multiple recipients. It can also be defined as a set of custom data fields retrieved solely for a particular user, presented as if the data was hand-entered and laid out, with embedded signatures, personalized for the recipient.

For the fax broadcast service bureau, issues like T.30 compatibility, pages per port per hour, and system fault tolerance are critical. In addition, compatibility with fax machines, fax cards, PDAs, and fax PADS are important concerns. Everyone has sent a fax that has failed due to line errors, and has successfully resent the fax. However, for a service bureau, even a single failure out of a thousand attempts is cost prohibitive. The most expensive minute of a phone call is the first, and each retry only escalates costs. Broadcasting 100,000 pages a day, a service bureau cannot afford 100 failures a day, or 3000 retries a month.

Effective port capability is important to the success of a fax broadcast service bureau. The ability to keep all fax ports configured and active is a necessity. "Pages per Port per Hour" measures the success of a computer based fax device, and is an important indicator of cash flow for a service bureau. Pages per Port per Hour can be affected by the software architecture, the fax device, and the fax device's subsystem.

There are many Computer Based Facsimile (CBF) devices on the market today, but GammaLink's product line is uniquely suited for fax broadcasting. GammaLink products have the highest compatibility rate in the world today. The firmware used to manage the T.30 connection was developed in 1985, and has been refined and enhanced for 10 years. Due to the unique GammaLink design of an Intel CPU "per fax channel," the impact of multiple fax devices in a host chassis is minimal; a 386/33 host PC is sufficient to drive a span of 24 GammaLink fax channels. GammaLink provides a ready-made fax subsystem optimized for delivering fax jobs to fax channels. The subsystem consists of a driver (the Dispatcher) and a shared "queue" file, allowing the broadcast developer to concentrate on document creation and inter-node communication; channel management and load balancing are handled by the GammaLink subsystem.

Application Examples

The most common Computer Based Fax (CBF) application today is in the delivery of mortgage rate sheets from commercial to retail lender. This typifies a computer based fax broadcast application. It consists of a standard sheet of data containing time critical information. The data must be delivered to the retail lenders in a window of opportunity, generally before 11:00 am or noon; otherwise, the data becomes dated and therefore obsolete.

Another popular user of fax broadcasting services is the travel industry. Imagine a cruise ship preparing to sail in three days, but the ship is only half booked. The cruise line might broadcast a special promotion to local travel agents advertising a "last minute deal" in an attempt to fill the boat. This formula also can be applied to charter plane flights, destination resorts, expeditions, etc.

What if you're a subscriber to the Wall Street Journal, and selected articles pertaining to your area of interest are faxed to your office. Or to manage your portfolio, at the end of each day, your custom stock and bond holdings are faxed over, detailing your portfolio's value. Any business which deals in time sensitive data and requires instantaneous delivery is a candidate for fax broadcasting.

The Node

The node is a computer where the fax cards reside, and the fax software runs. The system architecture, inter-connections, and file locations are designed into the node. The node generally runs the fax card driver, the node control process, a conversion routine, and a inter-node communications protocol. Any or all of these can be defined by the application developer. Of the four parts to the node, interaction with the other nodes and the control processes is the most important aspect, and should be carefully considered by future node developers.

Boards And Configurations

There are many fax boards which deliver fax documents 24 hrs per day, seven days per week, without failure or errors. GammaLink is regarded as the leader in high density PC-based fax products which offer a full range of densities, from single port to 6 ports per card.

137

Early fax broadcast systems were developed around the industry standard GammaLink CP card. In 1988, the systems were based on four channels per host PC (an Intel 80286). GammaLink was the first company to introduce T1 span support in a single host PC with the CP4/AEB fax card, this innovation allowed 24 fax channels, located in a single host chassis, to access a T1 trunk via a Dialogic DTI-124 header card.

Besides it extensive product line, available in the first quarter of 1995, will be GammaLink's CP12/SC card, a 12 channel card on a single slot which connects to the Central Office via the industry standard Dialogic SCbus digital telephony bus interface. The CP12/SC will support the latest T.30 specifications. In addition, it supports: V.17 modulation at 14.4 kbps, on the fly conversion to and from TIFF G3 (modified Huffman, MH), G3-2D (modified READ, MR) and G4 (modified modified READ, MMR), ASCII and PCX. The CP12/SC was designed with mission-critical fax broadcasting capability in mind. In two 16 bit ISA slots, a 24 channel fax broadcast node is created. The CP12/SC is a modular card, which can be ordered in a 6 channel configuration, to provide E1 span support in three slots. As with all SCBus compatible products, PEB bus support is integrated.

The CP12 product will be configured to support the industry standard MVIP digital telephony bus. In this environment, the CP12/MVIP operates as a resource module only, and does not provide switching functionality on the MVIP bus. Generally, this limitation is not an issue because timeslot switching is usually provided by the (telephone) network interface card.

The CP4/SC product offered by GammaLink continues to be offered for lower density fax applications, or systems in which 24 channels of fax are not required. Because of GammaLink's switching handler library, it is possible to dynamically allocate fax channels as resources. The CP4/SC supports the Dialogic PEB telephony bus interface as well.

Because of GammaLink's modular design, each fax channel operates independently. Therefore, if a single channel on the 12-port card were to fail, the other 11 ports would continue to run, and the work load would be automatically redistributed. Another advantage to the modular design is that the developer is presented with a single, unified interface to one or many fax channels. A system configured for 12 channels can be upgraded to 24 or 30 channels without source code modifications.

The Host PC

The host PC may have a major, or minor role, depending on the architecture of the broadcast system. Minimally, the host will be responsible for driving the installed fax devices, and possibly the (telephone) network interface card. The host will also maintain a connection to a main control processor which dispatchs fax jobs. In addition, the host may be responsible for performing file conversion, data merges, or running queries against remote databases. Note: Do everything possible to eliminate every aspect of the host PC from bottle-necking the performance of the fax system.

Because the fax broadcast environment is 24 hours a day, seven days a week, an industrial chassis is essential. Forced air cooling, redundant power supplies and sturdy rack mount design are all part of the requirements. Industrial chassis generally come with a passive ISA backplane which comes with a bus, and a "plug in" CPU card. The CPU cards tend to be all inclusive, containing a CPU, cache memory, extended memory slots, FDD and IDE HDD controllers. If the architecture calls for all the files to be stored locally, discard the IDE controller on the CPU card and replace it with a SCSI controller. Caching is a necessity, and can be handled by software via SMARTDRV, or the drive controller. More than one megabyte will not improve performance greatly on a SCSI controller; anything more than four meg will not be utilized effectively with SMARTDRV.

Generally, customers want the fastest CPU available on the market without considering other, more important aspects of a host PC. With a one megabyte caching SCSI controller, one really can't go wrong. If there are large amounts of data stored locally, or the node is configured for receiving, it might be advisable to run two smaller hard drives, and split the work load between them.

The quality of the video controller is less critical. Any $30 model will do, after all, a well designed system will provide remote diagnostics on a central control node. Files not critical for recording event, but frequently accessed by the system may be placed on a RAM drive for efficiency.

Most nodes today rely heavily on the LAN network for interconnection, requiring a fast Network Interface Card (NIC). The 100 mbps token ring is a powerful tool to reduce network lag times. With a FDDI backbone, the throughput is tremendous, and will effectively eliminate the LAN as a suspect in the elusive search for the bottleneck component of the node. Otherwise, build in monitoring tools to alert the system administrator of impending overloads on the network.

It is extremely important to realize that with data stored in relatively volatile RAM, such as RAMDrive or SmartDRV, a power failure is catastrophic. Therefore, Un-interruptable Power Supplies (UPSs) are a requirement. It is also advisable to maintain redundant lists of jobs active per node to be re-processed quickly if a node failure occurs.

Network Interfaces

Network interface implies connection to the Central Office (CO) or a local PBX to the node. In the US, the choices are limited for fax broadcasting; a 1.544 mbps T1 connection is most common. The current stage of ISDN Primary Rate Interface service in the US is not a viable option today. DS2 and DS3 (T3) trunks are multiple T1s bundled together. T1 trunks offer both DNIS and ANI through a wink start signaling protocol. There are many companies which provide network header cards; Dialogic, Mitel, Rhetorex, and Natural MicroSystems to name a few. Dialogic provides two products, the standard DTI/240SC, which is a T1 to SCBus connection or the D240SC/T1 which provides T1 termination and 24 channels of voice on a single slot.

In Europe and Asia, the G.703 standard, or E1 as it's known in the US, is "the digital trunk option." E1 is a 2.048 mbps digital line. The two popular types of service, Channel Associated Signaling (CAS), or an ISDN protocol, both of which have different flavors, or signalling protocols, depending on the country. E1 trunk header cards are primarily available from Dialogic and Acculabs. Dialogic's DTI 212 provides both CAS and ISDN trunk access and approvals in the large market countries around the world. Acculabs provides a widely approved dual E1 trunk access product for Europe with a PEB/SC or MVIP bus interface.

For most configurations, a Channel Service Unit (CSU) will be required. A CSU performs many functions, including terminating a digital trunk at the demarcation line, regenerating the signal and forwarding it to the PBX or network interface card. A CSU provides two primary activities; first, it ensures a clean signal is received into the final termination point, generally in the host PC; and second, a CSU can provide translation from the different types of signaling protocols available.

The largest concern facing US customers moving from T1 to E1 is that "all" line signalling in E1 is handled on timeslot 16. In T1, each timeslot has the ability to take it's channel off-hook with the A and B bits. With E1, all the line control information is sent via timeslot 16; for both CAS and ISDN protocols. Intelligent system architecture will allow easy migration between T1 and E1.

Software

There are three major aspects of the software, the users' interface, the administrators' interface, and the broadcast management software. The user interface software may be run at the service bureau site, but more commonly will run on the customer premise. This allows the customer to manipulate and edit broadcast lists, configure jobs, and edit data structures. The administrative software must be able to view customer configurations, as well as manage all the broadcast nodes. It must provide statistics about node usage and load balancing as well as warnings about impending failures. Finally, the broadcast management software is the underlying code that keeps the system running together. It is the operating system and the network.

The User Interface

The user's interface may be public, or strictly for internal use. Users of the broadcast service need to be able to update files, make changes to the broadcast phone lists, check status, etc. It might be that all users call a "Customer Service" line, and a representative provides them the information/changes required over the phone. This personalized attention may not be practical with the high cost of providing customer service reps. With today's computer literate users, access via modems or fax is the most practical solution.

Whether the broadcast system caters to internal "customers" through the main office of a corporation, or is a service sold to third party users, the most efficient way to update data sheets and phone lists is to have the user make their own changes. The interface could be a dial-up service, where the users get to add or subtract numbers from a "mailing list", or create new lists. They could check job status and completion times, get billing information and check phone charges. Obviously, providing dial-up service to manage accounts requires a significant programming effort, but provides the most flexibility and minimal impact on the service bureau personnel. The interface of choice is Windows, with it's GUI and world-wide acceptance.

For third party services, it would be easier to provide a "fax update" interface. Each service subscriber could have a special form, with a bar code logo which would be OMR'ed when received at the service bureau. This would allow for easy and automatic identification of the customer. There could be forms for adding numbers to a broadcast list, checking the numbers of a list, and any other service. If the user wished to update the broadcast document, it could be faxed in from a PC at the customer's site using ECM, which would provide a flawless reproduction of the image for re-broadcast. The user could send in a form to initiate a broadcast to a certain list, and request the completed information be faxed back upon completion. The return fax could contain all relevant information, including number of pages sent, completion time, and even an invoice for billing.

142

The Administrator's Interface

The administrator's software interface needs to be flexible and complete. Because the administrator needs access to every aspect of the system. A well defined interface allows instant statistics per node, number of jobs sent, pages per port per hour, percentage of time not utilized, error rates, etc. It should also provide a central monitoring function where ports or chassis with unusual statistics are reported. For example, channel 15 on node 241 is getting all errors, or node 93 is failing a more than an average number of jobs.

Along with the shear volume of number crunching, there are operational aspects to consider. Such as the ability to terminate a particular fax job. If the ABC Corp calls up and needs something immediately, there must to be a way to purge the system of pending jobs for a single customer. Difficulties arise with a distributed network where nodes bid for jobs based on Least Cost Routing algorithms. A second consideration is the ability to re-prioritize jobs. The largest customer calls to say they need list X broadcast in the next 20 minutes, there needs to be a way to replace the current job with a newer, higher priority job ASAP.

The Broadcast Management Software

The Broadcast Management software is responsible for controlling the node PC and communicating with a master control process. This is the most critical component of the fax system. It provides node management, some conversion routines, building documents, requesting new fax jobs, and returning fax jobs to the main controller. For the operating system, there is no substitute for a 32 bit flat model preemptive multitasking environment. NOTE: This is not Windows 3.1. It is OS/2 2.x, Windows NT, or Unix. The preemptive multitasking environment guarantees processes and/or threads operating on the CPU of service in a reasonable amount of time and for a finite amount of time. This assures each fax channel running in the chassis will be serviced and not be easily "starved" for data. Data starvation, or a buffer underrun condition, is much more common in DOS or Windows 3.1, where service times can not be verified.

143

It also enables effective communications between the node PC and the control PC, or even the node and a mainframe host which periodically downloads data.

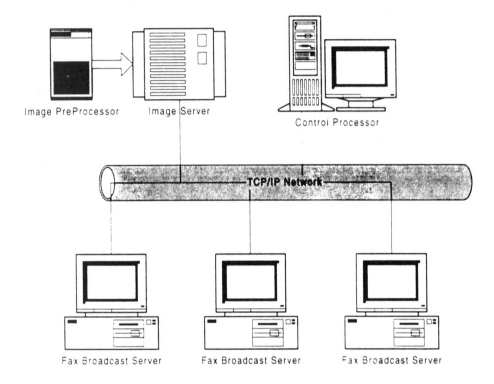

Figure 1: Sample fax broadcast network configuration

The Document and Document Creation

Fax broadcasts can be composed of nearly any type of data. Documents fall into two categories: static or dynamic data forms. Static data consists of forms or images that are fixed through the broadcast. This is the concept of "one to many". Such as single form transmitted to as many fax enabled devices as possible. The forms are easily updated once or twice a day. This process is ideal for the mortgage rate sheets. These forms are relatively static data and don't require much processing power to generate. They are generally updated daily, and the same data can be sent to all the recipients.

144

A static data broadcast can also be fax advertisements. In this case, the broadcast customer may create a one or two page fax ad to be sent to his customer list. This ad file can be created on any page layout program, such as PageMaker. In which a professional look and feel of desktop publishing. The file can be output as a PostScript document and then converted into the standard TIFF file format. It is important to note, Postscript to TIFF conversion routines are CPU intensive. It is not realistic to expect a host PC to convert PS files while broadcasting on 24 or more channel simultaneously. But in the "one to many" broadcast model, the original PS file needs to be converted once to be sent to thousands of sites.

Dynamic data forms are more complex, but parallel customer demands more closely. Dynamic data forms can be acquired from any database source using any query method. For example a Sybase database server or a FoxPro engine on a LAN, can generate custom reports if the data can be accessed quickly. This typifies the "many to many broadcast paradigm a different file for each phone number and illustrates the concept of personalized portfolio management by fax. Every day a new "report" is generated for each customer on the broadcast list. Each customer has a specified amount of information requested which is queried against a data source.

The GammaLink fax card is uniquely able to handle the custom data associated with dynamic forms and reports. GammaLink has an extensive ASCII to TIFF conversion routine that runs "on the fax channel," not effecting the host CPU. GammaLink supports up to four different downloadable bitmap fonts per channel, as well as page layout commands. It is possible to compress horizontally and vertically, and to change the resolution dynamically. There are also options to format the text per transaction.

Command	Parameter	Value Range	Default	Desc
Vertical Res	-VRx	0,1	0	Standard Res
Top Margin	-MTx	0-100	3	
Bottom Mgn	-MBx	0-100	3	
Left Margin	-LMx	0-50	14	
Right Margin	-Rmx	96-144	96	
Font	-Chx	0-3	1	Default Font
Page Length	-PL	0-999	66	Std Page Length
Char Height	-CH	8-16	16	Full Size
Char Width	-CW	8-16	16	Full Size

GammaLink's default bitmap font contains the standard line drawing characters which allow basic ASCII reports to be generated without an impact on the host's CPU. This allows generic data retrieved from a data server to be presented in a concise page layout, and facilitates creating custom data forms quickly, easily with minimal processing power.

GammaLink also provides a utility to create custom fonts for specific uses. This is helpful when migrating broadcast facilities to include overseas customers, where ASCII characters between 128 and 255 are loosely defined. This also allows users to create custom type faces to give standard characters a new look.

Document Input

Getting the document or data into the system for broadcasting is tied very closely to the user's interface. There are several methods for getting the final document to the service bureau to be broadcast. As mentioned above, it is possible to have the customer create the document in their office and data transmit it to the broadcaster in a page description language like PCL or PostScript. The major advantage of this, is that the customer creates the document exactly as they would like it. There is little impact on the service bureau; the image is converted to TIFF and it's ready to transmit. There is no loss of image in transit, and the customer knows exactly what the document will look like when it is faxed. The disadvantage is that the customer is required to invest in a software package and create the layout, or rely on third party desktop publishers.

Once the page is completed, there are several methods for entering it into the service bureau's system. The file could be uploaded to a BBS, or mailed through a public mail system such as CompuServe or MCIMail. Another alternative is to use a third party software package, such as Delrina's WinFaxPro with a standard 14.4 kbps fax/data modem. Delrina's package allows users to create a page in any software package and "print" this file to fax. This is similar to printing the file to PostScript and uploading the PS file, except the PS stepped is skipped, and Delrina converts the Windows metafile format directly into a faxable TIFF file. The user then faxes this file into a DID number to the service bureau which relies on the Error Correcting Mode of the fax device to ensure an exact replication of the file from the Delrina software. The DID number allows each customer to fax in a new data sheet daily. The service bureau can easily retrieve these files and automatically update the customer's broadcast list with the new file for the day. It would even be possible for the customer to specify on the cover sheet which broadcast list this file belonged too. Through an OMR process, the cover sheet could be scanned and a "broadcast list #2" might be marked, instructing the service bureau to update list number 2 and begin a broadcast on that list.

Another simple way to deliver broadcast data to the service bureau is using the new BFT protocol available on all computer based fax devices. BFT allows files to be data transferred to other CBF devices using existing T.30 protocol. It also allows the data to be transferred at 14.4 kbps without actually converting the data to TIFF. Generally files are smaller and send quicker. In addition, BFT has many secondary advantages such as automatic updating of end user software. Suppose the service bureau recently upgraded its end user interface. They could BFT broadcast the latest version of this interface to each of their end users. These service bureau explaining which file to execute could then send a fax to the end users to update the interface.

Desktop publishing is another business opportunity related to the fax broadcast service. Many firms will see the power of fax as a delivery mechanism for time sensitive data, but will not have the resources to generate professional data sheets. A full service fax bureau could maintain a desktop publishing department, or simply provide the service to it's subscribers.

Customization

Many fax users consider "customization" the act of sending a cover sheet with the recipients name and company, along with the "return address". Although this works fine when sending one or two faxes to other offices, consider the cost of broadcasting an additional sheet of paper to 10,000 users. Suddenly that cute cover sheet with the fancy logo becomes a huge cost overhead, and does little more than address the fax to a particular person. There are many parts to a fax broadcast that can be quickly and easily customized, without incurring a large overhead. Whether it be unique "header" lines, embedded TIFF images, or individual signatures, GammaLink's fax system makes it easy to create personalized documents without preprocessing the image or lengthy conversions.

The "header" of a fax is defined as the single line of text placed across the top of each and every fax page. The header can be generated by the sender (more common) or the receiver, or both. The header's function is to provide the receiver with information about where the fax came from and how to contact the sender in case of unsolicited faxes. Headers can also contain information about the date, time and number of pages in the fax job. GammaLink's newest line of fax devices all contain the ability to define a custom header on a per transaction basis. This feature is known as the "programmable header". It means each fax sent on each channel can be configured individually, to provide everything from different fonts, to different formats and even different times. There are two fields within fax transaction object which allow specific data to be associated with a particular job. This allows the recipient's name and company to be pasted across the fax as they receive it. This is accomplished with little more than a string copy put into the appropriate fields.

Programmable Headers

GammaLink's programmable header is just that, a single line of data at the top of each page, containing data elements that can be manipulated and customized on a per transaction basis. The programmable header contains the ability to place static elements (the originator's company name and return fax number) and dynamic arguments (the recipient's name and company) on each transaction.

Variable	Field Width	Description
&HEADER	20	HEADER field in queue rec
&USER	32	USER_ID field in queue rec
&FROM	20	Sent CSID, in queue rec
&TO	20	CSID to which the fax is sent
&PHONE	20	PHONE_NO field in queue rec
&PAGE	3	Current page number
&12HR	2	12 hr time 00 to 12
&24HR	2	24 hr time 00 to 23
&AMPH	2	Insert AM or PM value
&MIN	2	Minutes in 00 to 59
&SEC	2	Seconds in 00 to 59
&DATE	9	Date MM/DD/YY, format depends on Country Code
&TIME	9	Current time HH:MM:SS, format on Country Code
&ABMON	3	Abbreviation of month, "Jan, "Feb",...
&MONTH	2	Month in 01 to 12
&STRMON	10	Month in "January", "February",...
&DAY	2	Day of month in 01 to 31
&YEAR	2	Year in 00 to 9
&LYEAR	4	Year in 1900 to 2000

So using the above table, one could create the following header to appear on the top of a particular fax:

```
ABC Co Fax To: &USER at &PHONE    &DATE &TIME  Page &PAGE
```

Then to create a transaction object to customize the header:

```
strcpy (q_rec->USER_ID , "Robert Green");
q_rec->transmit_control |= GFQUSER_HEADER;
strcpy (q_rec->fn_send, "C:\\FILES\\GREEN\\FAX.TIF");
strcpy (q_rec->phone_no , "1-408-744-1900");
```

Note the value of "Robert Green" will be placed in the header of the document when it is sent, along with the phone number field, the date, time, and page number.

GammaLink also provides two additional undocumented features to it's programmable headers which are specific to fax service bureaus. First, is the ability to set the time shown on the header, not as the time where the fax originated, but of its final destination. This allows faxes sent internationally, or sent to a "store and forward" mailbox, to have the transaction time recorded in the local time zone, and not the originating Time Zone. This is simple, yet important feature to provide the feel of a "custom" fax in a broadcast environment..

The second feature is the ability to overlay a header over an existing header. Generally, headers are appended to the top of a fax page so they do not intrude upon the data. Because fax "store and forward" is so popular, service bureaus often rebroadcast images received from other machines which may contain headers from previous transactions. The overlay allows a new header to be imposed over (ANDed) the existing document or to integrated (ORed) over it.

Embedded TIFF Images

GammaLink makes it extremely easy to spruce up drab ASCII reports by allowing TIFF images to be embedded into the document and converted on the fly by the fax channel. One can create any type of image, and insert it into a standard ASCII file in TIFF format. The TIFF data can be a unique logo, image or even a signature file as a trailer.

Page Description Languages

Although it is quite easy to build a fax broadcasting business based on ASCII files with embedded TIFF images, the highest quality documents are created in desktop publishing software packages, and then "printed" to fax via PostScript or PCL page description languages. GammaLink provides an additional software product called GammaPage which contains both FAXPS.EXE and FAXPCL.EXE, which convert from PostScript or PCL to TIFF format.

Page Description Languages have the advantage of getting the WYSIWYF effect, or What You See Is What You Fax. It is important to note, however, that the majority of fax devices existing in the world today use "fine" resolution, which is approximately 200x200 dot per inch. This resolution is non-existent in printer world today; 600 dpi is becoming the standard. The T.30 specification does have provisions for a new "super fine" resolution, at approximately 400 dpi, but this is only useful if existing machines provide compatibility. Note that high resolution translates into more bits to send, and higher phone charges.

PDLs offer unparalleled image clarity and resolution, but there is a price. Conversion times are generally measured on the order of tens of seconds. The more complex the image, the greater the render time. On a standard "one to many" fax broadcast, this is not vital; the image is converted before broadcast, and the resulting TIFF file is stored on a file server to be accessed by each channel. However, what if there are 10,000 different PS files to accompany the 10,000 phone numbers of the broadcast list. A 24 line chassis, converting 24 PS files, at a rate of 10 seconds per image, or 240 seconds to put a fax job on each channel. When it takes 60 seconds to send a job, there is no way to keep all the fax channels busy.

To avoid this potential bottleneck, there is a company called Eicon Technologies in Quebec which manufactures a hardware assisted conversion utility. Basically, the PS or PCL file is downloaded to the conversion board so the image can be computed and rendered without using the host CPU. The rendered image is then compressed into TIFF and returned to the host system. The whole process takes 1-3 seconds.

Another possible solution is to set up an image server; a machine dedicated to converting images of any type. With enough memory and CPU speed, it would be simple to have a single RISC type machine keeping many fax channels busy. The converted files could then be stored in a file server for later retrieval. This machine could serve as an image preprocessor, doing customizations, ASCII formatting, and high speed image rendering.

151

Queuing And Traffic Dynamics

How Many Ports Do I Need?

"So, how many ports do I need?" is initially the most often asked question of fax broadcasters, and its not that simple. Understanding the queuing and traffic dynamics of your fax broadcast system are tantamount to implementing a successful service bureau. These types of statistics must also be built into the administration of the system to identify and avoid potential bottlenecks and log jams.

Pages per (PPH) is a critical phrase to understanding the success of a service bureau; however it is totally dependent on internal factors, such as document size, and external factors, such as busy receive fax devices. While it is possible to minimize document size with a few simple steps, sending to busy fax machines, sometimes as high as 40% during peak calling periods, become s a part of the dynamics of your fax traffic.

The net result of PPH is how many bits are being transmitted into the phone network. Most service bureaus bill by the minute, not the job or page, so they are not anxious to implement cutting - edge technology. It is important to remember that a fax device is really nothing more than a data modem with a standard protocol. Therefore, the two critical factors for all data transmission, the speed of the bits, and the number of bits (compression), are the gating factors. The majority of fax devices installed worldwide maintain the minimum support of 9.6 kbps and standard Modified Huffman (MH) image compression. New devices come with 14.4 kbps support, and better compression types like Modified READ (MR) and Modified Modified READ (MMR), offering 20% to 40% better compression over MH. GammaLink fax devices support MH, MR and MMR "on the fly," and will negotiate the best compression for use in the msp automatically.

Flowcharting the work as it passes through the bureau, and understanding the bandwidth of each component is critical. Documents entering the system as ASCII files require little or no processing, and can be quickly passed to the node for transmission. Documents which require data from outside database servers may or may not process quickly, depending on the server's work load. PostScript images will require preprocessing, but may be only one time. It may be required to flowchart different customers and their jobs as each will probably have a fairly standard impact on the system.

The number of ports installed may seem sufficient at the start, but as the service grows, or as customers put time constraints on broadcasts, growth becomes geometric, and the ability to scale the system will be tested. Customers who have time critical data will require the service bureau to deliver that data before a certain hour, or within a certain period. Suddenly, the number of ports required is not for the total number of pages per day, but must be able to handle peak load times. There may be 4 broadcasts of 10,000 jobs which must be completed before 10:00 am, with nothing to broadcast in the afternoon

Communication to the Nodes

Communication to the nodes is defined as how the fax node receives data and returns data to a controlling processor. There are several different approaches, each with it's own merit. First, GammaLink does provide it's own queuing option, which will be discussed more extensively late in this chapter. Therefore, one could put shared queue files and image data onto a common file server, and allow each node to access the file server independently. This is the simplest, least efficient method. It requires file sharing as the premise for exchanging data between nodes and the control process. It also requires each channel to read and from a common file on the network. Therefore, a single node will generate 24 requests for data from each channel. A standard ethernet connection would not sustain much growth.

A more efficient mode is a distributed approach, where each node communicates via a TCP/IP protocol. This would allow the control process to bundle a list of 100 phone numbers, and a file, and send it to a certain node. When the node finished processing,, it would return statistics on the 100 numbers, and place a request for more jobs.   The distributed approach is redundant.   As each node operates independently, any node could fail and the others would continue.   This eliminates a file server as a possible single point of failure.

And in large systems, there is no need for the nodes to be in the same building. An intelligent control node can route jobs to nodes in cities where the faxes are being sent, thus eliminating costly long distance toll charges.   "Every fax should be a local call" is a goal of many service bureaus. Also note, that at different times of the day, sending a fax from New York to Los Angeles is cheaper than from LA to NY.   This becomes more of an "odd by true" issues internationally.   Sometimes it is more expensive to send from Paris to Stockholm than to transfer the job to the US and send it from NY to Stockholm.

The network topology can effect the throughput of the system.   For smaller systems, ethernet will hold up fine. It is a simple and cost effective solution. But as your service grows, the requirements will increase on the network. Token ring is a solid platform, and provides the critical function: in a heavily loaded system, each node is guaranteed timely service. This allows the load to balance more evenly across all nodes, and to ensure each node gets access to the control processor and/or file servers.

Document Storage

The location of documents can adversely effect the throughput of both the installed fax channels and the network transport layer. Putting all documents on a single file server sets up a single point of failure, not to mention a single point of bottle-necking. It is important to realize that a network file server like Netware is designed and optimized to provide access to may files concurrently. Also note, if a single file is request ed regularly, that file never leaves the RAM on the file server and can be accessed in micro-seconds, not the milliseconds of physical partition.

The one drawback of a central file server is that it can introduce extraordinary amounts of network traffic. Each fax channel of a 24 line chassis will request the same data which needs to be passed over the network. Downloading the file to the node is probably most efficient. There is the obvious problem of having a different document for each fax job, in which case 24 documents would be downloaded as 24 jobs. This would generate roughly the same amount of network traffic as a single job on a common server. Therefore, to handle unique documents per job, a file server is the only real solution.

Building A Node

Hardware

Through out this chapter, we have recommended several different pieces to develop a node for a fax broadcast server. Here we put all the pieces together. First, there is the chassis. We recommend any one of the rack mount chassis offered by the Enhanced Platform Group (EPG) from Dialogic Corporation. We especially like any one of the Dialogic Telco Platform (DTP) chassis. Again, not to limit the node via the host CPU, a Pentium based CPU card with 256k cache running at 60 or 90 mhz will eliminate any potential bottlenecks. It also provide CPU power to spare when the time comes to increase the number of channels in a the node. A SCSI interface card is a reliable and reasonably priced product. If the system architecture calls for the data to be stored locally, and a RAM drive is not used, then investigate controller cards with on board cache. We have found 1 Megabyte to be sufficient, any more and performance does not dramatically improve. As for a network interface, SMC seems to provide a very wide range of OS compatibility. This is important for testing and system verification, but most likely, there will be one host OS, and therefore, whichever NIC you choose, it must run on that OS. SMC provides verified compatibility with DOS, OS/2 2.x, Interactive and SCO Unix.

GammaLink provides a full line of fax devices, and depending on the number of slots in the chassis, type of configuration, and anticipated

155

growth, different products might be better suited for different systems. A standard, 24 line fax broadcast server can use six of GammaLink's CP4/AEB cards connected to a Dialogic DTI-124. This is tried and true technology, easy to install ,and easy to maintain. If used with a "hot swap" chassis, problematic broads can be replaced without effecting the other channels in the system or having to power off, because each card is connected to the T1 trunk via an analog connection.

For more scalable systems, the CP4/SC provides compatibility with both the PEB and SCSA bus interface. This card is also available as a CP4/MVIP to provide an interface into the MVIP world. Natural MircoSystems provides a DTI-48, a dual T1 trunk interface via the MVIP bus,which is Ideal for integrating GammaLink fax into new or existing MVIP applications.

GammaLink's most recent introduction, the CP6/SC and (available in first quarter of 1995) CP12/SC, are the flagships of the product line. These high density cards are ideal for the fax platform environment. Providing a T1 trunk interface in two slots, or an E1 connection in three, the CP12/SC provides the density and scalability required in today's competitive fax platform market. Used in conjunction with the Dialogic DTI/211 for PEB, or the latest DTI/240SC, these products provide maximum flexibility while maintaining a unified API.

Software

The most critical part of the entire fax broadcast service is the software used to drive the installed fax channels on the node. The software is also used to provide feedback to the controller process, and provide information about jobs to the fax channels. First and foremost, an operating system needs be selected. Both OS/2 2.x and Unix (any flavor) are excellent choices. Both operating systems provide the ability to run a process per channel, and allow threads to be spawned to manage less critical tasks like updating the control process or downloading data from a file server/data server. Windows NT, scheduled for shipment in Q4 1994, is another excellent choice.

With it's TCP stack built into the OS, Windows NT provides distributed processing and peer to peer connectivity similar to many Unix flavors.

As a transportation layer, TCP/IP is the most efficient protocol, and provides the most flexibility today. It is MSP recommended for it's ability to run a peer to peer environment, thus enabling the distributed processing required for a successful service bureau.

The actual software architecture of the node is dependant on the level of control required by the application. A control process is required on the node. It has the responsibility for communicating with the master control process and with the fax channels. Communicating with the fax channels can be handled via a single entry point, or a process per channel.

As a quick overview, the GammaLink system is based on queue records, or fax transaction objects. The most basic fax transaction object contains a file name and a phone number. A queue record can also contain the retry strategy, delay between retries, header information, cover page, security features and other customizable data about a specific fax transaction. A queue record must be created for each job. The queue record is then entered into the GammaLink fax sub system via GammaLink's API (the GPI) to be processed. When a job is completed, it is returned to the application with all the relevant information pertaining to the transaction written into the queue record. There are two methods for getting a queue record to a fax channel, the queue interface and the interactive interface.

Queue Interface

The first is a high level, queuing interface, providing a single entry point into the fax sub-system. This is by far the most efficient and simplest interface to implement. The entire system is built on a shared file, called a queue file. The queue file is accessed by both the controller process running on the host and a GammaLink process called a job "dispatcher." The dispatcher hides all of the difficult, and sometime tedious development of a channel management system, including job buffering and load balancing. The dispatcher operates between the fax channels and the shared queue file. It periodically polls the queue file for new jobs and holds them in buffers in its memory space. As soon as a channel completes a job, the dispatcher hands it another. There is not more than a few milliseconds of lag time between jobs. This ensures that each channel is never idle, or put mildly, not generating revenue. The dispatcher also has the ability to balance the load on the fax channels. If a channel were to fail, the dispatcher would automatically re-route work to the other channels. Conversely, if more channels were added, the dispatcher could service them without <u>any</u> software modifications.

To enter a job into the fax subsystem via the queue file:

```
/*****************************************************************************/
Copyright (c) 1989-1992 by GammaLink, Sunnyvale, CA,
All rights reserved.

Purpose:     This code fragment demonstrates how to initialize and submit a record
             to the queue file.
/*****************************************************************************/

#include <stdlib.h>
#include <stdio.h>
#include <string.h>
#include "gfq.h"
#include "gfqpath.h"

int main(void)
{
        GFQRECORD      qrec; /* queue record data structure */
        char           qfile[GFQFILENAME_SIZE]; /* string for queue file */
        int            result; /* result code from function call */

    /* Get the location of the queue file     */

        if (gfqSearch (GFQDIR_QUEUE,"gfax.$qu",qfile)) {
           printf ("Error: Getting path to queue file\n");
           exit(1);
           }

        gfqClearRec(&qrec); /* Initialize Q Record data structure */

        qrec.operation = GFQDIAL_SEND;        /* Operation to be done */
        strcpy(qrec.phone_no,"1-408-744-1900");/* Phone number to call */
        strcpy(qrec.header,"Test Fax");       /* Programmable header field */
        strcpy(qrec.fn_send,"TEST001.TIF");          /* Document to send */

    /* Submit to into fax subsytem for immediate processing*/

        result = gfqSubmit(qfile, &qrec);
```

```
        if (result) {
        printf("Error:  Queue submit failed, result = %d\n",result);
        return(1);
        }
}
```

This demonstrates how to create a queue record, copy in transaction data, and place the queue record into the queue file. There are approximately 45 fields in the queue record, this shows the minimum needed to send a fax. Notice there are only two function calls required, the gfqClearRec ();, which NULLs out the data structure and sets some default values, and gfqSubmit (); which places the queue record into the queue file for processing.

```
/***************************************************************************/
Copyright (c) 1989-1992 by GammaLink, Sunnyvale, CA,
All rights reserved.

Purpose:      This code fragment demonstrates how to read queue records out
              of the queue file
/***************************************************************************/

#include <stdio.h>
#include <stdlib.h>
#include <string.h>

#include "gfq.h"
#include "gfqpath.h"

void main (void)
{
    GFQRECORD       qrec;        /* queue record struct */
    char            qfile[GFQFILENAME_SIZE]; /* queue file */
    char            filename[GFQFILENAME_SIZE]; /* send filename */
    char            channel [GFQFILENAME_SIZE]; /* channel name */
    int             status; /* status or fax job */
    int             result; /* result code from function call */

    /* Get the location of the queue file */

    if (gfqSearch (GFQDIR_QUEUE,"gfax.$qu",qfile)) {
    printf ("error finding queue file\n");
    exit(1);
    }

    /* Read the last (oldest) record in the Sent list */

    result = gfqFindFirst(qfile, &qrec, GFQSENT_LIST, GFQLIST_END, "");

    while (result == GFQSUCCESS) { /* Test the status */

        strcpy(filename, qrec.fn_send); /* read info from record */
        strcpy(channel, qrec.modem_id); /* fax channel number */
```

```
    status = qrec.status;   /* job status */

    printf("Found record in Sent list\n");
    printf ("Channel = %s\nStatus = %d\nFile = %s\n",
            channel, status,filename);

    /* get the next record */
    result = gfqFindNext(qfile, &qrec, GFQREAD_FWD, "");
    }
  printf ("GF Queue file status = %d\n", result);
}
```

This scalability and reliability are tantamount to providing peerless
support for mission critical fax broadcast applications.   The real
advantage to the queue file interface is in time to market. By managing
all of the lower level queuing and fax traffic interface, a broadcast
developer can concentrate on the user interface, or building better/faster
document customization routines. Fax broadcast jobs are, for the
most part, a batch process. Therefore, the fax transaction should
be treated as such, and basically submitted to a "fax processing system"
which sends the data, and returns a status.

Interactive Interface

Known internally as the "real time" interface, or the "named pipe"
interface, GammaLink's GFD library functions provide the lower
level control professional developers require. The model of the
Interactive interface is a process or thread per fax channel. Each
thread maintains a pipe or data stream to the fax channel. It is possible
to run a single pipe in duplex, or two separate pipes. The application
must create and manage each pipe. The pipes used are native to
the operating system, unless DOS is used, in which case the GFD
library emulates an OS/2 named pipe interface.

A message structure, called a datagram, is created to communicate
on the pipe.   A datagram is a set of data structures, each with a
generic message header:

```
struct gfx_rt_header {
      unsigned short source;          /* Source address    */
      unsigned short destination;     /* Destination address */
      unsigned short function;        /* Function number    */
```

```
        unsigned short status;              /* Returned status    */
        unsigned long  id;                  /* transaction id     */
        };
```

which is then incorporated in other structures to generate different message types:

```
struct gfx_rt_message {
        struct gfx_rt_header  header;       /* message header */
        char   info[2048];                  /* data buffer */
        };
```

The 2k data buffer is used to transport any type of data or messaging, primarily queue records.

The most common use of the Interactive interface is simply to submit a queue record to a fax channel via the named pipe, and to get a completed queue record in return when the job is completed.

```
/*********************************************************************/
Copyright (c) 1989-1992 by GammaLink, Sunnyvale, CA,

All rights reserved.

Purpose:      This code fragment demonstrates how to initialize and submit a record
              via a named pipe to a certain channel
/*********************************************************************/
#include <stdlib.h>
#include <stdio.h>
#include <string.h>

#include "gfd.h"
#include "gfqctl.h"

int main(void)
{
        GFQRECORD       qrec;       /* queue record data structure */
        int             result;     /* result code from function call */
        int             ret;        /* return code from Remote Procedure Call */
        int             channel;    /* channel to submit job to */
        int             chassis;    /* chassis to submit job to */

        gfqClearRec(&qrec);         /* Initialize Q Record data structure */

        qrec.operation = GFQDIAL_SEND;                  /* Operation to be done */
        strcpy(qrec.phone_no,"1-408-744-1900");         /* Phone number to call */
        strcpy(qrec.header,"Test Fax");                 /* Programmable header field */
        strcpy(qrec.fn_send,"TEST001.TIF");             /* Document to send */

    /* Submit to into fax subsytem for immediate processing*/
```

161

```
     result = gfqRemoteRequest(chassis,
                               channel,
                               GFXQRECORD,
                               0x0,0x0,
                               &qrec,
                               sizeof (qrec),
                               &ret);

 if (ret || result) {
    printf("Error:  Pipe submit failed, result = %d,
              return = %d\n",result, ret);
    return(1);
    }
}
```

Once the named pipes are in place, there are other functions which can be enabled. It is often required for developers to get real time status about the transaction during the fax call. For this, GammaLink provides an Event Notification interface. The T.30 protocol is a flow chart, and at the most requested locations, GammaLink has enabled the ability for the application to receive an event that a certain point has been reached in the protocol. When the event is received, there are a number of decisions to be made by the application. The most common choices are to continue with the call, or to abort. It is possible to collect DTMF digits, generate a tone, submit another queue record, or simply get status.

Event	Phase	Actions	Info	Time Outs
Call_Pending	pre Phase A	Continue Abort Queue Record Tone Digits	None	30 Sec
Answer	pre Phase A	Continue Abort Queue Record Tone Digits	DTMF digits	30 Sec
Info_Exchange	end Phase A	Continue Abort	DIS/CSI/NSF	3 Sec
Page_Break	end Phase C	none	Page Sent	0 Sec
Call_Term	Phase E	Continue Abort	Completed Queue Record	30 Sec

For fax broadcasting, there are two powerful uses for the event feature. The first is security. Because of the Info_Exchange event, it is possible for the application to get the CSID of the remote machine. With this, it is possible for the application to decide if this is the correct machine, and continue or abort the data transmission. This is mostly popular with the Law, Medical and Banking fields, due to confidentiality of the data. The second is with the page break notification. Because the T.30 protocol timings are very tight around page turn-around, this is a notification, without the ability of action. With this event, it is possible to provide a real time status of jobs to a user remotely. Simply having the application count the page breaks will provide the user/administrator of an immediate and exact picture of the fax channel's current state.

The Future

The next generate fax broadcast servers will probably be dual purpose. During the daylight hours, they will be providing fax on demand and LAN fax services. And during the night, they will be broadcasting. So dynamic allocation will be a must, as well as the ability to integrate with FOD and LAN fax products.

Broadcast servers may be required to provide both fax and data broadcasts. The new BFT specification contains a simple method for exchanging both data and fax. This will allow both data and images to be delivered to a user's desktop. BFT is also important because it allows companies to instantly and automatically distribute the latest version of software.

Beyond fax and data, it is subject to question, some say video, others say integrated messaging, voice, text and fax. Whatever the bits, and whatever the transport mode, the net result is the same, delivering time critical information to a user instantly, and the new Dialogic SCSA is capable of integrating all the technologies seemlessly.

Section VI

Building a Two-Call Fax-On-Demand System Using GammaLink Equipment and Ibex Technologies Software

Building Two-Call Fax-On-Demand System Using GammaLink Equipment and Ibex Technologies Software

Introduction

Ibex Technologies, Inc. (550 Main St, Suite G, Placerville, CA 95667, 916-621-4342,) sells Microsoft Windows based fax-on-demand systems and has been in the market since 1989. They have been extremely successful in the high technology market so if you have ever requested an automated fax from one of the large software companies, you have probably used an Ibex system.

Ibex's strengths are the Windows interface and the ease-of-use of the software for both the system administrator and the system operators who add documents, examine reports or modify the announcements or applications. Because of the leadership position and time in the market, Ibex has been able to create and market ancillary FOD products such as automated forms processing, Lotus Notes integrations, fax server, high volume fax broadcasting, host interfaces, and other products.

Two-Call vs. One-Call

The Ibex system supports two types of fax delivery modes: "One-Call" and "Two-Call". One-Call mode means that the caller must call from their fax machine, and the resulting fax transmission is sent back on the same telephone line. Callers can call a Two-Call system from any phone - the system will prompt the caller for a fax telephone number to be used to send the fax. Its called Two-Call (or Call Back on some systems) because the system will deliver the fax with a second call. You may mix One-Call and Two-Call modes on the same system. For example, you may wish to designate that some incoming lines operate in One-Call mode, while others operate in Two-Call mode, or you may even specify that a voice line may operate in One-Call and Two-Call modes depending on the time of day, who the caller is, or some other criterion.

One-Call Fax Delivery Mode

An important aspect of systems set up in One-Call mode is that the system throughput will not be as high as a similar system (same number of voice and fax ports) set up in Two-Call mode. This is because a Two-Call system can have all voice ports and all fax ports busy at the same time. A One-Call system must have fax ports idly standing by whenever a voice call is in process. In fact, most One-Call systems including Ibex's physically connect each voice port to a fax port, thus guaranteeing that there will be a fax port available when the voice port requires one. Other important aspects of the One-Call method include:

• Phone usage costs are lower since the fax transmission takes place on the caller's initial phone call. This needs to be weighed against having to purchase more voice and fax ports.

• Useful for international callers, since they do not have to enter their international phone number as the system would have to dial it from the host country.

• Callers must call from a fax machine in order to receive a fax.

• The phone line is tied up for the duration of the fax transmission. No new phone calls can come in on that voice telephone port.

Two-Call Fax Delivery Mode

The Two-Call method requests a fax telephone number from a caller after a fax has been chosen. The fax transmission is sent immediately if no other faxes are waiting, or it is queued if all available fax boards are busy. With this method, FactsLine will process incoming calls at the same time that it faxes out requested information. Thus it is possible (with a four voice, four fax line system) to have four callers listening to voice menus and making choices while four previous callers are receiving faxes. Other important aspects of Two-Call are:

• More convenient for callers since they do not have to be at a fax machine to use the system.

• Can be used to send faxes to someone other than the caller.

• Costs more for the FOD system owner in that all fax transmissions are separate telephone calls initiated by the FOD system.

entered by callers. This is especially true for international callers who must know to enter the country code, city code and phone number to successfully receive a fax.

Although a significant percentage of FOD systems used One-Call FOD during the beginning of the FOD market in the late 80's, most Ibex systems installed today are Two-Call. In fact, one of the largest and oldest FOD applications is currently switching from One-Call to Two-Call. However it is still used where cost is important and also for international applications. The remainder of this section concentrates on Two-Call fax delivery.

Sizing a System

The size of a FOD system is measured in ports, and is usually broken down further into voice ports and fax ports. For example, an Ibex 4x4 system contains four voice and four fax ports. An average FOD system has the same number of voice ports and fax ports (average voice call is two minutes and the average number of pages is 2-3, or 2-3 minutes of fax time). It is common, however, to require more fax ports than voice ports if the faxes sent tend to be lengthy and/or you plan to implement fax broadcasting as well. It is less common to have more voice ports than fax ports; this is usually seen in applications where IVR (Interactive Voice Response) features are used which can lengthen the caller's time on the voice port.

The following table sizes a typical FOD system that experiences an average call time of two minutes. To use the table, first look up your average fax time across the top of the table (use one minute for one page, unless you know you are going to be faxing lots of graphics or use high resolution, then use 1.5-2 minutes per page). Then look up the maximum, peak number of document requests you expect in an hour along the left hand side of the table. The table will then provide the number of voice and fax channels required. Sizing a One-Call system is quite different - this table sizes only a Two-Call system.

Fax Time in Minutes

Requests per hour	2	3	4	5	6	7	8	Voice Ports
	NUMBER OF FAX CHANNELS REQUIRED							
5	1	2	3	3	4	5	5	2
10	2	2	3	4	5	5	6	3
15	2	3	3	4	5	6	7	4
20	2	3	4	5	6	7	7	4
25	3	4	4	5	6	7	8	5
30	2	3	4	6	7	8	9	5
35	2	4	5	6	7	8	9	6
40	3	4	5	6	8	9	10	6
45	3	4	5	7	8	9	11	6
50	3	4	6	7	9	10	11	7
55	3	5	6	8	9	11	12	7
60	3	5	6	8	10	11	13	8
65	3	5	7	8	10	12	13	8
70	4	5	7	9	11	12	14	8
75	4	6	7	9	11	13	15	9
80	4	6	8	10	12	14	15	9
85	4	6	8	10	12	14	16	9
90	4	6	8	11	13	15	17	10
95	4	7	9	11	13	15	17	10
100	5	7	9	11	14	16	18	10

Table Courtesy of FaxMax and Ibex Technologies, Inc.

Assumptions for Two-Call system sizing:
Fax time based on exponential distribution
Five percent of fax requests delayed five minutes
Formula: #channels = (fax time /60) * (36 + requests / hour)
70% of incoming voice calls result in fax request
Incoming voice calls last two minutes (mean duration)
One percent of all callers get a busy
Formula: Erlang loss tables

A Real-World Application: Symantec

Symantec Corp., developers of applications and system software such as Norton Utilities, Q&A and ACT, installed an Ibex system in 1991 after experiencing a telephone support crisis. The company released two upgrades of key products. Call volume jumped from 20,000 phone requests to 50,000 in one month. Callers experienced longer than average waiting periods and the company was loosing calls, due to hang-ups. Simultaneously, customer service representatives were manually faxing dozens of product information sheets per day.

A reseller of Ibex systems, FaxMax, installed a 24 port system at Symantec in 1991 using Ibex's older DOS based software with Dialogic D/41 four port voice boards and GammaLink XPs one-port fax boards. The system was upgraded and expanded by FaxMax in 1994 using Ibex's Windows based software and GammaLink CP4LSI four port fax boards. The computers are racked AST 386 and 486 PCs on a Novell 10baseT Ethernet network, supported by an uninterruptable power supply and tape backup system. The racked FOD equipment is connected to the company's corporate network using a bridge router. Ibex's Interactive Forms software was also added in order to automate the processing of forms that were faxed into the system. The system is maintained by a project coordinator, and requires approximately two hours per week of attention for installing and testing new files, archiving and backing up, and transaction report generation.

Symantec reports, to date, the system has sent over 750,000 faxes.

 **Automated Fax System**

800-554-4403
Product Information Available 24 Hours Per Day, 7 Days Per Week . _ S and Canada Onl.

Symantec Customer Service

TO: The person at extension JACK

FROM: Symantec Customer Service

YOU REQUESTED THE FOLLOWING DOCUMENTS:
110   Q&A for Windows Data Sheet
750   The Norton Desktop for Windows Data Sheet

Symantec's Facsimile Retrieval System
Custom Designed by FaxMax (415) 965-4553   Ibex Technologies Software (916) 621-4342

Symantec Service And Support Center
175 West Broadway / Eugene, Oregon / 97401

Customer Service (General Product Information, Orders and Returns) Phone: 800-441-7234 / 503-334-6054

Boards and Configurations

Ibex systems use Dialogic voice and GammaLink fax boards. Dialogic two-port and GammaLink one-port boards can be used for small systems, while Dialogic high-density boards and GammaLink four-port boards can be used for larger systems. A system can be set up in a standalone (non-networked) configuration using a 386 or better computer supporting up to 24 ports of voice and fax.

In order to be able to add documents from other computers in real-time or modify applications on-the-fly, the Ibex system must be installed on a network. Novell, Microsoft Windows NT, LAN Manager or other full service networks are recommended. Peer-to-peer networks such as Lantastic or Windows for Workgroups may work for small applications, but are not recommended because of the high network traffic that the voice and fax applications can produce.

Telephone Interface

Typical FOD installations use analog lines either straight from the telephone company (POTS), or from the PBX. If the lines are from the PBX, they need to be analog lines (like those used for a fax machine or modem) and not the digital or proprietary lines. Either way you will want your voice lines to roll over to the next line available line (called a hunt group). If you have a large system with more than one computer with voice lines in them, you will want to alternate the roll over lines between the computers to build in redundancy.

Dialogic T1 interface boards (or AcuLab ISDN boards for Europe) are used for large capacity systems. In this case one computer usually supports 24 ports of voice or fax, and multiple networked computer chassis are used to create the system. DID/DNIS, and ANI are supported and can be used to enhance the application.

Since a Two-Call system has separate voice and fax lines, they can be treated completely separately. For example, the incoming voice lines can be routed through your PBX and auto-attendant, and the fax lines go straight out to the central office. Or you can use a T1 card with the voice lines, but use analog lines for the fax ports.

Application Setup

Most IVR and FOD systems use a scripting language to generate applications. Ibex systems break this tradition and uses a data-driven architecture. A Windows front-end provides an easy-to-use interface to modify the application. This design allows the application or document storage to be changed in real-time without taking the system off-line. The record-locking, multi-user aspect of the databases allow for contention and simultaneous changes.

Voice Menu

The core of an Ibex application is the Voice Menu, a voice announcement that allows callers to perform actions by pressing a key on their touch-tone phone. The Ibex configuration software allows you to configure a voice menu within Windows.

The right hand side of the Voice Menu represents the telephone keypad. For example, right now as currently configured with this Voice Menu, if the caller presses key one, they will be sent a fax as configured in Fax Box named "INDEX". If the caller presses key two, they will be sent a fax as configured in Fax Box "DOCNUM" (this Fax Box actually allows the caller to enter document numbers). If the caller presses key zero on their telephone, they will be transferred to an operator.

To assign an action to a key, you would just click with the mouse pointer on the key you wished to configure. A list of action options will then pop up. The available actions from a Voice Menu are:

- Go to another Voice Menu
- Send a Fax
- Record a message (simple voice mail)
- Message retrieval
- Collect data (zip code, account number, etc.)
- Call transfer
- Schedule actions based on time of day

175

- Host Interface option
- Interactive voice response option
- Credit card processing option

Once the caller actions are configured for a particular Voice Menu, a voice announcement needs to be recorded which presents the options to the caller. These can be studio recorded and imported, or they can be recorded on the spot during the application creation. The Ibex software supports either recording direct to the Dialogic voice board (call from a telephone), or you may record the announcements using a SoundBlaster compatible sound board. To initiation a recording, you would click on the "Announcement" button in the Voice Menu. This brings up the voice recorder, plus the written script for this announcement.

Once the announcement is recorded, the Voice Menu is ready to go. For more advanced applications, the "Options" menu of the Voice Menu allows access to password options, multiple languages, logic (IF-THEN statements), and other options.

Fax Selection

After the Voice Menu(s) is created, you need to define how callers can request faxes. For a simple application you select how many faxes the caller may receive, the length of the document number and the cover page. For more advanced applications, you have the option to choose from the following:

- Which prompts to play, or to customize prompts
- Limit access to specific document group
- Change header text depending on various factors
- Use different cover page depending on various factors
- Use fax number contained in caller account database
- Use binary file transfer instead of fax transmission
- Change number of fax attempts or retry delay
- Activate credit card processing

176

Document Management

One of the most time consuming tasks with any FOD system is the addition, deletion and management of fax documents and the index of available documents (although not required, many FOD systems have an index of available documents that is available to callers). Ibex systems provide a lot of value in this area by offering Windows print drivers, document management tools and automated indexing tools.

The following document types are directly supported by Ibex (no manual pre-conversion necessary).

- ASCII Files
- TIFF Group 3 / F (Fax TIFF)
- TIFF Modified Group 3
- TIFF Group 4
- PCX
- FMT (Cover page format file)
- FDL (Fax Description Language Files)
- Lotus Notes Documents
- Wang OPEN/image documents

The images are managed with the following screen from the Ibex FactsLine configurator. The Description and Category fields, for example, are used to create the automatic index of available documents. The Description can also appear on the cover page, if you enable that feature.

If the "Accounting" button is depressed, then you may assign an owner (supplier) of the document, a meter value which will automatically count and limit the number of access, and the value of the document for credit card charging or for reporting purposes.

If the "Schedule" button is depressed, then you may schedule the new document (or the document modification) to be activated at a later date. Or you can schedule a document to be expired at a certain date.

177

Adding Documents to an Ibex System

Faxing-in

If the document is not available in electronic format, the easiest method is to fax the document into the system. This can be done remotely from a different site (i.e. New York to San Francisco) or it can be done locally with the keyboard. The image quality will be the same as if you manually faxed the document to the caller - i.e. not the best quality that can be obtained.

Scanning

Scanning is similar to faxing in documents with a slight increase in the quality of documents. You would expect a significant quality increase when scanning, and you can get great results with the right techniques, but with most scanning software doing a "one-pass" approach you will get about the quality of a faxed document.

Importing and Conversions

The Ibex Viewer program that views fax images also converts images from one format to another. Most programs in Windows, DOS or the Macintosh can produce a image in .PCX, TIFF, or .BMP format, which the Ibex supports. There are also third party conversion programs, like Hijaak, that support more file formats.

Precision Fax Service

Extremely high quality fax document creation is more of an art than a science and no software package exists that can consistently produce high quality documents from a variety of sources. Ibex offers a service, Ibex Graphics, that uses the necessary dithering, filtering and rendering techniques to create truly stunning fax documents.

Easy Does It.

Q&A for Windows gives you several options for viewing your data, including the familiar row-and-column format of a spreadsheet and customized layouts to match any form you're already comfortable with. Add tools that give you extremely easy access to your data, along with built-in templates and on-line tutorials, and no matter how you look at it, Q&A for Windows ensures that you won't have to learn a completely new way of working.

Put DAVE To Work For You.

Q&A for Windows incorporates a feature called DAVE (Do Anything Very Easily), which makes it easy to accomplish a wide variety of tasks—from accessing information in databases using everyday English phrases, to creating simple, yet powerful "scripts" that automate often-repeated actions. The result is that you spend less time working on the structure of the database, and more time working with your data.

Create Dazzling Documents.

To create professional-looking documents, Q&A for Windows includes a full-featured word processor that's integrated right into the program. With Q&A's word processor, you can create exciting documents using big, bold headlines and a wide variety of type styles. You can even incorporate great-looking charts, tables, and graphs in your documents—adding more graphic appeal to your projects.

Customize Your Mailings

Q&A for Windows lets you generate personalized mailings using names, addresses, and other information from

different database programs such as Q&A, dBASE and Paradox. There's even a "sort and retrieve" function so you can select entries from your master mailing list based on various criteria.

Generate Comprehensive Reports.

Q&A for Windows gives you extensive reporting capabilities with free-form and columnar reports. And Q&A's multi-pass report writer lets you build the most incredibly precise, finely-tuned reports ever.

You're Already Compatible.

You don't have to reinvent the wheel when you need to use different applications. Q&A for Windows makes it easy for you to share data with other software programs such as dBASE, Paradox, 1-2-3, Excel, Word and WordPerfect. So now you can retrieve and incorporate information from other sources, quickly and easily.

Unite Your Workgroups.

Q&A for Windows not only allows you to easily share information with other popular software applications, it offers superior support for workgroups. Extensive network compatibility, the ability to transparently share information with Q&A for DOS, and a wide range of security features are among its powerful network capabilities.

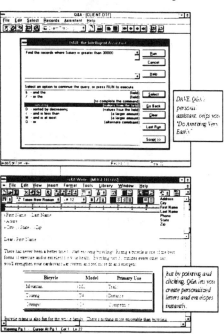

DAVE, Q&A's personal assistant, helps you "Do Anything Very Easily."

Just by pointing and clicking, Q&A lets you create personalized letters and envelopes instantly.

System Requirements

- IBM PC using 80386 or higher microprocessor, or 100% compatible
- 4 MB RAM
- VGA display adapter or higher resolution
- Hard disk drive, floppy drive
- DOS 3.3 or later
- Windows 3.1 or later, in enhanced mode
- Microsoft mouse or 100% compatible
- Networks compatible with Windows 3.1 or later

Symantec Corporation
10201 Torre Avenue
Cupertino, CA 95014-2132
800-441-7234
Fax 408-255-3344

Windows Print Driver

The Ibex Print Driver will create fax documents from any Windows program such as Word for Windows, WordPerfect or PageMaker. You would call up the document using your application, choose the Ibex Print Driver as your active printer, and print. If the workstation is on the same network as the Ibex System, then the print driver will communicate with the system and add the document to the FOD library in one step. In the process you can also specify the document management options (such as scheduling or pricing) as described above. If you are not on a network, then you may create a TIFF image file to be copied or carried to the Ibex System.

Macintosh Documents

Some Ibex customers use Postscript to fax conversion on Mac documents while some, since Ibex supports Acrobat files, use Adobe Acrobat to create the document on the Macintosh. And some prefer using their Mac fax modem to simply fax the document to the Ibex FactsLine system (not good for gray-scale images).

Lotus Notes Documents

Ibex supports Notes documents directly through the FactsLine for Notes product. Notes users can specify which Notes documents are to be included on the FOD system, and which Notes "Form" is to be used to create the actual fax document. The system will cache requests in order to minimize network traffic and reduce the load on the Lotus Notes server.

Wang OPEN/image

Many imaging systems use Wang OPEN/image as the core technology to store and index images. Ibex FactsLine can directly access the images, convert them to fax images "on-the-fly" and fax them to callers on request.

180

Index of Available Documents

FactsLine can automatically create an index of documents for callers. You can place all documents in one index, or create separate indexes using different groups of documents. The document description, document number, category, number of pages, and the date the document was last updated can be included in the index.

Advanced Topics

Most FOD systems require only the simplest of voice menus (indeed, the simpler the better), static non-changing documents and simple fax queuing and management. However as the industry grows and systems grow in size and importance, new features and capabilities are required.

Fax Description Language

High impact fax documents typically require a mix of constantly changing textual data and graphic images. For example, you may wish to merge a person's name onto a fax-on-demand or fax broadcast cover page using nice fonts and graphics. Or you may wish to create invoices or other forms from dynamic data.

Intelligent fax boards such as the GammaLink, Brooktrout and Intel fax boards will convert text to fax "on-the-fly" by using the on-board processor of the fax board. However, they convert the text to a fixed-size courier type font that is rather bland. Moreover, they cannot mix text and graphics in one location. They can send a horizontal band of graphics then switch to text for a portion of the page, but the result is a nice graphic document with glaring non-graphic text above or below it. Thus there is the "standard" fax-on-demand cover page look - a nice graphic header that drops into straight text.

Ibex's Fax Description Language (FDL) produces full-graphic cover pages and fax documents with unlimited combinations of font types, sizes and graphics. FDL allows you to mix text and graphic elements on a fax page by allowing you to describe the font, size, color (black, white or gray) and placement for the text elements and the placement of graphic elements. The text can be static text, or you may specify FactsLine variables that may be fax-on-demand caller information, broadcast information for fax-merge features, or data extracted from a database. You can also perform basic drawing functions like creating lines, boxes and circles on the page using various pens and brushes. For example, you may wish to draw a black box on the page with white italicized text word-wrapped inside the box.

Host Interfaces

Host Interface capability allows the caller to obtain information from another personal computer, a mini-computer, or a mainframe. The information is usually based on DTMF digits that the caller entered. These entries may be as simple as an account number, or as complicated as a series of variable length numbers the caller enters in response to prompts. The Ibex Host Interface option is flexible enough to be used with an extremely wide variety of information sources and information formats.

Fax Queuing

As systems grow in size, shaving a few seconds off each transmission and managing the fax resources efficiently can produce significant savings. The Ibex Transaction Manager (ITM) provides these features, and is designed for larger systems that use more than one computer to create a virtual FOD system. For example, if a fax transmission fails after two pages were sent, ITM will only send the remaining pages, not those already sent. ITM also implements an intelligent retry strategy by analyzing the information returned by the fax board, and will not retry the transmission if it determines the receiving end is not a valid fax device. ITM also supports email integration for user notification, fax server features, high volume fax broadcasting services, and transaction or entire-job management.

File-on-Demand

File-on-Demand is similar to fax-on-demand, except that it allows binary data files to be delivered to your fax modem instead of an image delivered to your fax machine. File-on-Demand feature combines the ease of use of fax-on-demand with the functions of a bulletin board system. Unlike a bulletin board system, File-on-Demand uses the global fax standard - no knowledge of communication parameters or logon procedures are needed. Ibex was the first fax-on-demand company to offer file-on-demand, called the Ibex FileBack feature.

FileBack opens up new possibilities for software support services and automated software distribution. Updates, bug fixes and diagnostic programs can be transmitted on demand to users, and software can be ordered and delivered within seconds. Include the Ibex Credit Card Module and callers can enter their credit card number for online credit card charging before the software is delivered.

Voice Recognition

Although not used in the USA, voice recognition (VR) is gaining popularity in Europe and South America because of the lack of touch-tone capabilities in many of these countries. FOD is an excellent application for VR because the voice application is quite simple and only number recognition is required. Speak a few numbers to select document(s), speak your fax number and you are done. There is usually no need to recognize letters or words, which is where VR starts to fail.

Service Bureau Applications

Service Bureaus own FOD equipment and charge others to use it. They typically push the system in areas that CPE (Customer Premise Equipment) customers do not. Ibex systems provide service bureaus with the ability to run large number of applications on the same

183

system (usually based on DID number), account number access with database tracking, on-line or batch credit card charging, and high volume broadcasting.

Section VII

**Fax Processing Examples Using
Brooktrout Technology Equipment**

Fax Processing Examples Using Brooktrout Technology Equipment

Introduction

If you've worked in any computer-related technology, like programming for example, you know that there's always more than one way to solve a problem. A good illustration of this, is the way different manufacturers design fax boards. Brooktrout has developed a high-performance, multichannel board and created the Universal Port by integrating fax and voice in one board. Brooktrout's API is designed to support fax functions simply as well as provide applications developers with great degree of control in setting up their applications to support unique requirements.

Brooktrout Product Line

Brooktrout offers multi-channel boards for advanced fax and integrated fax-and-voice systems. Brooktrout's TR Series fax and voice processing boards include the TR114ᴛᴍ Series 2-, 4-, and 8-channel Universal Portᴛᴍ boards and the TR112 Series 1- and 2-channel boards. These boards can be used with fax functions only or with integrated voice capabilities. All Brooktrout boards come with diagnostic test routines to simplify installation troubleshooting.

In addition, specific products offer combinations of the unique capabilities that are useful in certain applications:

> TR Series boards with integrated direct-inward-dialing (DID) interfaces simplify the development of systems using DID telephone service for routing of inbound faxes (for example, routing on a LAN or creation of a fax mailbox service) or selection of a specific application from several applications that run on a single system. These boards are available either with DID interfaces only or a combination of DID and loop-start interfaces.

> TR Series boards with MVIP, PEB, OR SCSA telephony interfaces can be used either in systems with digital telephone service (T1, E1, or ISDN) with digital interface boards or in systems using specialized resources; such as speech recognition, text-to-speech, or pulse detection, that run on separate boards.

> TR Series boards with loop-start interfaces approved by PTT's around the world permit applications developed on a TR Series platform to be installed in different countries.

The Universal Port

TR114 Series

The TR114 Series Universal Port boards eliminate the limitations imposed by single-function boards. Each Universal Port on the TR114 Series performs advanced fax and voice processing. Each channel increases system reliability and provides unequalled flexibility by supporting: independent state-of-the-art fax and data transmission and reception, voice recording and playback, DTMF detection and generation, and call processing.

The Brooktrout TR114 Series offers two to four channels per card and up to 48 channels per chassis. Each channel of a TR114 Series board has a dedicated Digital Signal Processor (DSP) and 32-bit CISC microprocessor and real-time operating system. Every TR114 port is supported by up to 2MB of DRAM and a custom Brooktrout ASIC, a discrete architecture providing unequalled flexibility and reliability. This configuration, unique to Brooktrout, ensures that each channel's function is totally independent and that all resources are available on every channel.

TR114 board support advanced fax features:

> 14.4 kbps transmission
> MMR, MR, MH compression
> Error-correction mode
> T.434 binary file transfer
> Microsoft At Work

TR114 Series boards are available in several configurations:

* 2- or 4-channel analog board with loop start or on-board DID telephone interface

* 2-, 4-, and 8-channel digital boards with MVIP or PEB interface for integration with digital network interfaces or resources from other vendors.

TR112 Series

For smaller applications or as an entry-level card, the Brooktrout TR112 offers developers of multi-channel fax systems low system implementation costs and great flexibility in choosing an operating system. The TR112 Series supports Group 3 fax features in one-and two-channel configurations, both of which are available with loop start telephone interfaces.

The TR112 Series 2-channel board also features Direct Inward Dialing, or DID, (for automatic distribution of faxes in shared fax system environments) or AEB interfaces, as well as speech playback and DTMF detection and generation (for systems requiring voice prompting and DTMF input). In addition, all Brooktrout TR Series fax cards are compatible, so there is a clear upgrade path.

The Brooktrout philosophy of product design, as realized in its products, has been accepted around the world. [show table of international certifications.]

Brooktrout API and Software Tools

Brooktrout's principal applications development tool is the Brooktrout Fax/Voice Applications Programming Interface (API). The API is a tool kit of C language functions, sample applications, and debugging tools which permits application programmers to generate high-performance multi-channel fax and voice applications under many different operating systems.

The API architecture is based on Brooktrout's layered fax/voice systems model:

APPLICATION (system-dependent) - the APPLICATION layer implements the specified system function.

BTMANAGER (system-dependent) - The BTMANAGER layer provides database and scheduling functions which

192

may vary widely depending upon system requirements. In many systems, these functions are performed by an independent processor connected via a LAN.

BTCALL - the BTCALL interface layer supervises inbound/outbound faxes and voice calls and data transfers. The T.30 facsimile protocol is easily accessed, and the need to understand every detail of its implementation is eliminated. Functions are included to play and record speech files. Programmers can easily establish and monitor calls, accept inbound routing, and record and play back speech.

BTBIOS - the BTBIOS layer provides an interface to the fax and voice line cards.

HARDWARE - The HARDWARE layer includes fax and voice processing hardware and firmware.

The Application and BTManager layers are the portion of the code developers would write; Brooktrout's API and hardware provide the BTCALL, BTBIOS, and hardware layers.

API's Multi-Processing Approach

The Brooktrout API was designed with a multi-processing approach to constructing applications and controlling the activities of the board. The basis of the design is the principle of using independent processes (or threads) in the BTCALL layer to control each fax (or voice) channel on the board. These processes are addressed by one or more separate processes in the Application layer and the BT Manager layer.

The BTCALL layer comprises independent processes for each channel in the system. These processes control all of the activities of the channel, such as sending or receiving a fax, playing a voice prompt, interpreting DTMF tones, etc.

The processes send information to the board by transferring commands and data to buffers in the BTBIOS layer. The Processes in the BTCALL layer receive interrupts generated by the board from the BTBIOS layer.

The BT Manager and Applications, such as managing the queue of outgoing faxes, filing received faxes, and notifying users of their receipt, or playing a fax-on-demand application, are written as separate processes that can access the BTCALL layer processes independently. This gives the application direct access to control of each individual channel.

This multi-tasking approach offers developers the opportunity to easily create better performing systems that can easily be scaled to handle larger volumes. Giving the application direct access to the processes controlling the board offers developers greater flexibility in creating systems that meet their requirements. The interrupt-driven approach to controlling the activities of each channel reduces the CPU utilization by eliminating the need for the application to constantly poll each channel to check its status. Instead, the board generates interrupts only when necessary, and the processes are run only when functions are required. Eliminating the fixed interval of an application- driven polling routine also operates the channels more efficiently and quickly, increasing the overall system throughput.

Multi-tasking systems are also easily scaled. Adding new channels simply requires replicating processes for the new channels, with minimal changes to the supervising manager or application processes. Brooktrout's API reliably supports up to 48 channels of fax and voice processing in a single server, with larger systems created in networked architectures.

Brooktrout supports most popular multi-tasking operating systems, including UNIX (SCO, Interactive, and others), OS/2 2.1, Windows NT, and QNX. Brooktrout has also developed a similar, pseudo-multitasking to creating applications under MS-DOS, which offers many of the same benefits, despite the fact that MS-DOS is not a multi-tasking operating system. In MS-DOS, the applications and BT Manager are incorporated in a foreground thread, which is the main thread of control. The BTCALL layer functionality is contained in background threads, one thread per channel.

Sample Applications

Sending and Receiving Faxes

Brooktrout provides three different levels of fax functions: high, middle, and low. Generally, the high-level functions simplify the process of sending and receiving a fax. The middle-level functions provide more flexibility and control than the corresponding high-level functions, but they require more knowledge of the basic steps involved in sending and receiving faxes. The low-level functions provide the greatest control over the fax process.

The high- and middle-level functions use the Brooktrout Infopkt data structure, while the low-level commands use raw data files only. An Infopkt consists of a tag and associates data. The Infopkt organizes the different data types (fax, ASCII, TIFF, PCX/DCX) into a single structure for either transmission or reception.

Brooktrout fax boards convert ASCII, TIFF, and PCX/DCX files to fax format on the fly, with no "preprocessing" by the application required. Infopkt or raw data files that are used to send faxes can include these data types, as well as standard Group 3 files (MH, MR, or MMR). The Brooktrout boards can also combine different data files on the fly during transmission to assemble a fax automatically from different data files, even within the same page.

The Brooktrout API includes sample code for sending and receiving faxes with high- and low-level commands, as well as fax polling. The following application demonstrates how to send a fax using the high-level fax functions. In this example, the local id is specified from the command line. The file btcall.cfg is the configuration file.

```
/*
 * Code fragment for Sending a Fax using high level API calls.
 * Brooktrout API entry points begin with Bfv...
 */

    /* Attach to a channel using BfvLineAttach. This function opens
the specified channel and initializes the channels line structure. A
pointer to the line structure is returned and then used with each
API call from then on */

    if ((lp = BfvLineAttach (chan_number) = = NULL)
    {
    fprintf (stderr, "can't attach to fax board\n");
    exit (1);
    }
```

```
/* BfvLineReset will reset the channel, read a configuration file,
and download channel information included firmware, fonts, and
country specific parameters */

    if (BfvLineReset (lp, "btcall.cfg") < 0)
    {
    fprintf (stderr, "Reset failed\n");
    exit (1);
    }
```

```
/* BfvInfopktOpen opens the infopkt stream file from disk for read
only. The function returns a pointer, ips, to s structure if successfully
opened. This ips will include all the necessary data and control
to send the fax. This can include: ASCII, G3/G4 image, TIFF,
and PCX data. It can also contain: automatic page header, resolution,
wide, maximum bit rate, and page break control information.

if ((ips = BfvInfopktOpen ("faxfile.ips", "r") = = NULL)
    {
    fprintf (stderr, "can't open %s\n", faxfile.ips);
    exit (1);
    }
```

```
memset((char *)&fres,0,sizeof(fres));   /* initialize the fax result
stucture, fres */
```

```
/* BfvLineOriginateCall dials the phone number specified and waits
for a final call progress value. The result is stored in the call result
structure, cres. The possible call progress values include: fax tone
```

detected, human detected, ring no answer, busy, special information tones (SIT) for invalid numbers, no dial tone. The result can be used to develop sophisticated retry strategies. Also, a user defined function can be passed to BfvLineOriginateCall for reporting of intermediate call progress results such as ringback, and answer supervision.*/

```
BfvLineOriginateCall (lp, "w449-9009", CALL_PROTOCOL_FAX,
0, &cres, 0, 0);
if (cres.res.status != BT_STATUS_OK || cres.res.line_status
!= FCP_ANSWER_TONE_DETECT)
    {
  printf ("Fax Tone not detected\n");
  printf ("Final Result was  %d", cres.res.line_status);
  exit (1);
    }
```

/* Fax tone (either CED or V.21 FSK) was detected */

/* Use BfvLoopCurrentDetectEnable to monitor loop current dropouts on the line. On MVIP and PEB cards, this will monitor the A signalling bit. This ensures channels are not off-hook longer than necessary */

```
BfvLoopCurrentDetectEnable(lp);
```

/* BfvFaxSend transmits one or more fax pages as specified in the ips. The fax identification (TSI) can be passed as an argument to be sent to the receiving fax machine. When the function returns, the fax result structure, fres, contains the success/failure status. If failure, the fres.res.line_status gives the exact reason for the failure. */

```
BfvFaxSend (lp, ips, "Brooktrout FAX", &fres);
if (fres.res.status != BT_STATUS_OK)
{
    printf ("Fax Sending Failed \n");
    printf ("Fax Send Error hangup code is %d \n", fres.res.line_status);
    exit (1);
}
```

/* loop down the page result list */

```
pageres = fres.reslist_head;
jj = 0;
while (pageres)
{
    printf("page number: %d\t", jj + 1);
    if (pageres- > bft_type)
    {
        static char *bft_names[] =
```

197

```
{NULL,"T.434","BTM",NULL,"DTM",NULL,NULL,NULL,"EDIFACT"};

        printf("bft type %s\n",bft_names[pageres->bft_type]);
    }
    else
    {
if (pageres->direction)
printf("no. ascii bytes %d\t",pageres->ascii_bytes);
printf("bad lines %d\ttotal_lines %d\n",
            pageres->bad_lines, pageres->total_lines);
    }
    temp = pageres;
    pageres = pageres->next;
    free(temp);
    jj++;
}
```

```
BfvInfopktClose (ips);

BfvLineDetach (lp); /* detach the fax channel so another process/thread
can use it */

}
```

The following segment demonstrates how to receive a fax into an infopkt format. The Brooktrout API offers the option of automatically storing received faxes in any format defined by the application. The board will convert the document to the preferred format on the fly (as it is received). This is particularly useful in minimizing the file dize of received documents by converting documents to MMR format (the most compressed file format available).

Playing and Recording Voice Files

Brooktrout's TR114 Series Universal Port boards also support voice processing functions. Like fax functions, there are three levels of speech functions. The high-level functions record and play raw speech files directly to and from a file. The mid-level functions record and play Infopkt-formatted speech files to and from an Infopkt stream. The low-level functions record and play raw speech files to and from a data buffer.

The following examples illustrate recording and playing back voice files using the high-level commands.

/* Code fragment for answering an inbound call, playing a speech file, and hanging up */

```
if ((lp = BfvLineAttach (chan_number)) = = NULL)
    {
fprintf (stderr, "can't attach to channel %d\n",unit);
exit (1);
    }


if (BfvLineReset (lp, "btcall.cfg") < 0)
    {
fprintf (stderr, "Reset failed.\n");
p_exit (1);
    }
```

/* Open a speech greeting file for reading */

```
if ((ips = BfvInfopktOpen ("greeting.ips", "r") = = NULL)
    {
fprintf (stderr, "can't open %s\n", name);
p_exit (1);
    }
```

/* BfvLineWaitForCall waits for an incoming call for the duration specified in the timeout. If timeout is
* zero, it waits forever.
*   An incoming call could be from an analog loop line, a DID line, or a digital line (e.g.
* T1). In the case of DID, DNIS, or ANI, the application can access the digits via the
* cres software structure.
*/

```
BfvLineWaitForCall (lp, timeout, &cres);
if (cres.res.line_status != WAIT_FOR_CALL_OK)
{
    printf ("BfvLineWaitForCall timed out \n");
    exit (1);
}
```

/* Use BfvLineAnswer take the channel off-hook */

```
BfvLineAnswer (lp);
```

/* Use BfvLoopCurrentDetectEnable to monitor loop current dropouts on the line. On MVIP and PEB cards, this will monitor the A

199

signalling bit. This ensures channels are not off-hook longer than necessary */

BfvLoopCurrentDetectEnable(lp);

```
    /* BfvSpeechPlay plays a speech file read from the ips file*/
BfvSpeechPlay (lp, ips, play_func, NULL, &res);
if (res.status != BT_STATUS_OK)
    {
        printf ("Error occured during speech playback \n");
        exit (1);
    }

printf ("Speech Playback completed successfully \n);

BfvInfopktClose (ips);

BfvLineTerminateCall (lp, &res);  /* hangup the line */
}
```

/* Code fragment for answering an inbound call, recording a speech file, and hanging up */

```
if ((lp = BfvLineAttach (chan_number)) == NULL)
    {
    fprintf (stderr, "can't attach to channel %d\n",unit);
    exit (1);
    }


if (BfvLineReset (lp, "btcall.cfg") < 0)
    {
    fprintf (stderr, "Reset failed.\n");
    p_exit (1);
    }
```

/* Open a speech recording file for writing */

```
if ((ips = BfvInfopktOpen ("message.ips", "w") == NULL)
    {
    fprintf (stderr, "can't open %s\n", name);
    p_exit (1);
    }
```

/* BfvLineWaitForCall waits for an incoming call for the duration specified in the timeout. If timeout is
* zero, it waits forever.
*   An incoming call could be from an analog loop line, a DID line, or a digital line (e.g.
* T1). In the case of DID, DNIS, or ANI, the application can

access the digits via the
* cres software structure.
*/

BfvLineWaitForCall (lp, timeout, &cres);
if (cres.res.line_status != WAIT_FOR_CALL_OK)
{
    printf ("BfvLineWaitForCall timed out \n");
    exit (1);
}

/* Use BfvLineAnswer take the channel off-hook */

BfvLineAnswer (lp);

/* Use BfvLoopCurrentDetectEnable to monitor loop current dropouts
on the line. On MVIP and PEB cards, this will monitor the A
signalling bit. This ensures channels are not off-hook longer than
necessary */

BfvLoopCurrentDetectEnable(lp);

/* Using BfvSpeechRecord, record incoming speech for 15 seconds
at 32 K bits/second ADPCM */

BfvSpeechRecord (lp, ips, oki_adpcm, RATE_8000, BITS_4, AFE_8000,
SPCH_MSB, 15000L, 0L, 1,   play_func,
NULL, &res);

if (res.status != BT_STATUS_OK)
{
    printf ("Error occurred during speech record \n");
    exit (1);
}

printf ("Speech Recoding completed successfully \n);

BfvInfopktClose (ips);
BfvLineTerminateCall (lp, &res); /* hangup the line */

Creating a Fax-On-Demand Application

The Brooktrout API makes it easy to create integrated
fax and voice applications using the capability of the
Brooktrout TR Series boards. The following fax-on-
demand application combines the fax and voice
capabilities illustrated earlier with DTMF reception
capability.

```
/*
 * Sample Brooktrout Fax and Voice Application Program
 * Copyright (c) 1989-1994 Brooktrout Technology Inc. All rights
reserved.
 */

#define FILE_GREETING "ivr_msg/greeting.pkt"
/* This speech prompt contains the following opening greeting:
   Press 1 to leave a message, 2 to get a message,
   3 to receive a fax, 4 to send a fax, and 9 to rerecord the greeting.
*/

#define FILE_BYE        "ivr_msg/bye.pkt"
/* This prompt says goodbye. */

#define FILE_LEAVE      "ivr_msg/leave.pkt"
/* This prompt indicates to leave a 15 second msg to be recorded.
   Recording will last for 15 seconds or until a touchtone is pressed.
*/

#define FILE_RECORD     "ivr_msg/record.pkt"
/* This prompt asks the user to begin a new greeting msg to be
recorded.
   Recording will last for 15 seconds or until a touchtone is pressed.
*/

#define FAX_XMIT        "ivr_msg/fax_xmit.pkt"
/* This is the fax infopkt file to transmit */

/* These are created by the application at runtime */
#define FILE_MSG        "ivr_msg/msg.pkt"
#define FILE_NEW        "ivr_msg/new.pkt"
#define FAX_RCV             "ivr_msg/fax_rcv.pkt"

/* Volume (gain) of beeps played */
#define BEEP_VOLUME -10


#include "btlib.h"

int unit;
int done;
int (*nodefunc) ARGS((BTLINE *));
int lflag;

/* usage is printed when user starts program with no options */
void
usage ()
{
    fprintf (stderr, "usage: ivr [-u unit] [-L]\n");
```

```
        exit (1);
}

void
main (argc, argv)
int argc;
char **argv;
{
    extern char *optarg;
    extern int optind;
    int c;
    BTLINE *lp;

    set_debug_mode(-1);  /* API-level debug output for real time
status display */

/* getopt parses command line options, "u" is the Brooktrout channel
number,
    "L" is whether is loop to the beginning when a call has completed
*/

    while ((c = getopt (argc, argv, "u:L")) != EOF)
    {
    switch (c)
    {
      case 'u':
        unit = atoi (optarg);
        break;
      case 'L':
        lflag = 1;
        break;
      default:
        usage ();
    }
    }

    if ((lp = BfvLineAttach (unit)) == NULL)
    {
    fprintf (stderr, "can't attach to unit %d\n", unit);
    exit (1);
    }

    if (LINE_TYPE(lp) != BOARD_TYPE_TR114)
    {
    fprintf (stderr, "Channel %d is not a TR114.\n",unit);
    exit (1);
    }

    dowork(lp);

}
```

```
void dowork(lp)
BTLINE *lp;
{
    for (;;)
    {
    BfvLineReset (lp,"btcall.cfg");

    process_call (lp);

    if (!lflag)
        exit(0);
    }
}

void
process_call (lp)
BTLINE *lp;
{
    CALL_RES cres;
    int timedout;
    int val;
    int (*oldnodefunc) ARGS((BTLINE *));
    struct infopkt_stream *ips;
    RES res;

    BfvLineWaitForCall (lp, 0L, &cres);
    if (cres.res.line_status != WAIT_FOR_CALL_OK )
    {
    BTERR bterr;

    BfvErrorMessage(lp,&cres.res,&bterr);
    printf ("BfvLineWaitForCall: %s\n",bterr.long_msg);
    exit (1);
    }
    BfvLineAnswer (lp);

/* BfvToneDetectEnable enabled the channel DTMF detection capability
*/

    BfvToneDetectEnable(lp,DTMF_16TONE);

    nodefunc = node_top;

    for (timedout = 0; timedout <= 3; )
    {
    oldnodefunc = nodefunc;
    val = (*nodefunc)(lp);
    if (val)
        break;
    if (oldnodefunc != nodefunc)
```

```
          timedout = 0;
      else
          timedout + +;
      }

      if (val > = 0)
      {
   ips = BfvInfopktOpen (FILE_BYE, "r");
   if (ips)
   {
      BfvSpeechPlay (lp, ips, NULL, NULL, &res);
      if (res.status ! = BT_STATUS_OK &&
          res.status ! = BT_STATUS_USER_TERMINATED)
      {
          BTERR bterr;

          BfvErrorMessage(lp,&res,&bterr);
          printf ("BfvSpeechPlay: %s\n",bterr.long_msg);
          exit(1);
      }
      BfvInfopktClose (ips);
   }
   else
   {
      fprintf (stderr, "can't open %s\n", FILE_BYE);
      exit(1);
   }
      }
}

/* tell play to stop as soon as any tone appears in the tone buffer
*/
int
play_func (lp, arg)
BTLINE *lp;
char *arg;
{
   if (BfvTonePeek (lp) < 0)
    return (0); /* keep going */
   else
   return (1); /* stop now */
}

/* play the speech data in the named file, stopping if a tone comes
in.
*/
void
play_msg_stop_on_tone (lp,name)
BTLINE *lp;
char *name;
{
```

205

```
        struct infopkt_stream *ips;
        RES res;

        ips = BfvInfopktOpen (name, "r");

        if (ips)
        {
        BfvToneFlush(lp);
        BfvSpeechPlay (lp, ips, play_func, NULL, &res);
        if (res.status != BT_STATUS_OK &&
            res.status != BT_STATUS_USER_TERMINATED)
        {
            BTERR bterr;

            BfvErrorMessage(lp,&res,&bterr);
            printf ("BfvSpeechPlay: %s\n",bterr.long_msg);
            exit(1);
        }
        BfvInfopktClose (ips);
        }
        else
        {
        fprintf (stderr, "can't open %s\n", name);
        exit(1);
        }
}

/* play a message, but don't stop if a tone comes in */
void
play_msg (lp,name)
BTLINE *lp;
char *name;
{
        struct infopkt_stream *ips;
        RES res;

        ips = BfvInfopktOpen (name, "r");

        if (ips)
        {
        BfvSpeechPlay (lp, ips, NULL, NULL, &res);
        if (res.status != BT_STATUS_OK)
        {
            BTERR bterr;

            BfvErrorMessage(lp,&res,&bterr);
            printf ("BfvSpeechPlay: %s\n",bterr.long_msg);
            exit(1);
        }
        BfvInfopktClose (ips);
        }
```

```
        else
        {
        fprintf (stderr, "can't open %s\n", name);
        exit(1);
        }
}

/* each node is represented by a function. When the function is
 * called, it plays the node message, then takes node specific
 * action. For most nodes, the message lists the menu of choices,
 * and the rest of the action is to wait for a tone. If no
 * tone arrives in the node specific timeout, the function just
 * returns. The higher level code calls the function again to
 * make it repeat the message and do its action again. The
 * higher level code is responsible for deciding whether we have
 * stopped making progress and to hang up. If a tone does arrive,
 * the node uses it to select a new node, and sets the function pointer
 * "nodefunc" to the new function, then returns. The higher level
 * loop then calls the new nodefunc, which will play the new message,
 * etc. Other actions include starting up the fax board, etc.
 */

/* this is the top level node */
int
node_top (lp)
BTLINE *lp;
{
    int digit;
    RES res;

    /* this message says:
     *
     * Press 1 to leave a message, press 2 to get the message,
     * press 3 to send a fax, press 4 to receive a fax,
     * press 9 to record a new greeting.
     */
    play_msg_stop_on_tone (lp, FILE_GREETING);

    /* wait up to 5 seconds for a tone, then repeat message */
    digit = BfvToneGet (lp, 5000L, &res);
    if (digit < 0)
    {
    if (res.status != BT_STATUS_TIMEOUT)
    {
        BTERR bterr;

        BfvErrorMessage(lp,&res,&bterr);
        printf ("BfvToneGet: %s\n",bterr.long_msg);
        exit(1);
    }
    return (0);
```

```
        }
        switch (digit)
        {
           case '1':
nodefunc = node_leave_message;
break;
           case '2':
nodefunc = node_get_message;
break;
           case '3':
/* Receive a fax means ivr program transmits */
nodefunc = node_xmit_fax;
break;
           case '4':
/* Transmit a fax means ivr program receives */
nodefunc = node_rcv_fax;
break;
           case '9':
nodefunc = node_new_greeting;
break;
           case '*':
return (1);
        }
        return (0);
}

int
node_new_greeting (lp)
BTLINE *lp;
{
        struct infopkt_stream *ips;
        RES res;

        play_msg_stop_on_tone (lp, FILE_RECORD);

        if ((ips = BfvInfopktOpen (FILE_NEW, "w")) == NULL)
        {
fprintf (stderr, "can't create %s\n",FILE_NEW);
exit (1);
        }

        BfvTonePlayBeep( lp, TONEID_CCITT_1, 100L, BEEP_VOLUME);
        printf ("Recording...\n");
        BfvToneFlush(lp);
        BfvSpeechRecord (lp, ips, CODE_ADPCM, RATE_8000, BITS_4,
AFE_8000,
                SPCH_MSB, 15000L, 1000L, 1, play_func, NULL,
&res);
        if (res.status != BT_STATUS_OK &&
        res.status != BT_STATUS_USER_TERMINATED)
```

```
    {
BTERR bterr;

BfvErrorMessage(lp,&res,&bterr);
printf ("BfvSpeechRecord: %s\n",bterr.long_msg);
exit(1);
    }

    BfvToneFlush(lp);

    BfvInfopktClose (ips);

    remove (FILE_GREETING);
    rename (FILE_NEW,FILE_GREETING);

    nodefunc = node_top;

    return (0);
}


int
node_leave_message (lp)
BTLINE *lp;
{
    struct infopkt_stream *ips;
    RES res;

    play_msg_stop_on_tone (lp, FILE_LEAVE);

    if ((ips = BfvInfopktOpen (FILE_NEW, "w")) == NULL)
    {
fprintf (stderr, "can't create %s\n",FILE_NEW);
exit (1);
    }

  BfvTonePlayBeep( lp, TONEID_CCITT_1, 100L, BEEP_VOLUME);
    printf ("Recording...\n");
    BfvToneFlush(lp);
    BfvSpeechRecord (lp, ips, CODE_ADPCM, RATE_8000, BITS_4,
AFE_8000,
            SPCH_MSB, 15000L, 1000L, 1, play_func, NULL,
&res);
    if (res.status != BT_STATUS_OK &&
    res.status != BT_STATUS_USER_TERMINATED)
    {
BTERR bterr;

BfvErrorMessage(lp,&res,&bterr);
printf ("BfvSpeechRecord: %s\n",bterr.long_msg);
exit(1);
```

```
        }

        BfvToneFlush(lp);

        BfvInfopktClose (ips);

        remove (FILE_MSG);
        rename (FILE_NEW,FILE_MSG);

        nodefunc = node_top;

        return (0);
}


int
node_get_message (lp)
BTLINE *lp;
{
        play_msg (lp, FILE_MSG);
        nodefunc = node_top;
        return (0);
}

int
node_rcv_fax (lp)
BTLINE *lp;
{
        struct infopkt_stream *ips = NULL;
        FAX_RES rres;

        nodefunc = node_top;

        BfvTonePlayBeep( lp, TONEID_CCITT_1, 100L, BEEP_VOLUME);

        if ((ips = BfvInfopktOpen (FAX_RCV, "w")) == NULL)
        {
        fprintf (stderr, "can't create %s\n",FAX_RCV);
        exit (1);
        }

        BfvFaxReceive (lp, ips, "this is the local id", &rres);
        if (rres.res.status != BT_STATUS_OK)
        {
        BTERR bterr;

        BfvErrorMessage(lp,&rres.res,&bterr);
        printf ("BfvFaxReceive: %s\n",bterr.long_msg);
        }

        BfvInfopktClose (ips);
```

```
    return (-1);
}

int
node_xmit_fax (lp)
BTLINE *lp;
{
    struct infopkt_stream *ips = NULL;
    FAX_RES sres;
    CALL_RES call_res;

    nodefunc = node_top;

    BfvTonePlayBeep( lp, TONEID_CCITT_1, 100L, BEEP_VOLUME);

    if ((ips = BfvInfopktOpen (FAX_XMIT, "r")) == NULL)
    {
    fprintf (stderr, "can't open %s\n",FAX_XMIT);
    exit (1);
    }


BfvLineOriginateCall(lp,"",CALL_PROTOCOL_FAX,0,&call_
res,NULL,NULL);
    if (call_res.res.status != BT_STATUS_OK ||
    call_res.res.line_status != FCP_ANSWER_TONE_DETECT)
    {
    BTERR bterr;

    BfvErrorMessage(lp,&call_res.res,&bterr);
    printf ("BfvLineOriginateCall: %s\n",bterr.long_msg);
    goto x_done;
    }

    BfvFaxSend (lp, ips, "this is the local id", &sres);
    if (sres.res.status != BT_STATUS_OK)
    {
    BTERR bterr;

    BfvErrorMessage(lp,&sres.res,&bterr);
    printf ("BfvFaxSend: %s\n",bterr.long_msg);
    }

x_done:

    BfvInfopktClose (ips);
    return (-1);
}
```

Section VIII

Faxing Internationally

Faxing Internationally

Introduction

Facsimile technology and its widespread use is not an American phenomenon. It is a valuable business tool, in use worldwide. As such, the system integrator and user doing business internationally, must be certain that those fax devices and fax systems, that are vital links to customers (and their customers) across the globe, not only work as promised, but are fully compatible electronically and legally with the public telephone networks of the countries where these systems will operate. The only way to ensure one hundred percent compatibility with the public telephone networks of other countries, is by having the main components of these systems, in this case CBF boards, officially approved, or homologated by the government authorities in charge of public communications in each country, (generally the Post Telephone & Telegraph (PTTs) administrations.)

When a system integrator (and customers using the system) uses a CBF board that has not been homologated, by the country where it will operate, the risk is run that subtle electronic incompatibilities between the unapproved CBF board and the foreign public telephone network may crash the system and even damage the public network. Less serious results can be higher telephone bills due to incompatible signal transmit levels, receiver sensitivities, and other factors that may cause the system to do early disconnects (wasting calls), or prolonging calls by forcing it to fall back to considerably slower transmission speeds.

Not every CBF manufacturer, however, pursues homologations. The process is expensive, time-consuming, and requires inordinate patience. Some manufacturers, have recognized that because fax is a universal technology customers must have solutions that cut across international borders. This requires that the manufacturer create and maintain a credible presence in the world market. Before this can become a reality, however, it becomes necessary to comply with the different requirements and specifications for communication products, deemed essential by each nation.

Without these international approvals no manufacturer can assure customers that the systems they create and use will work in the countries where they operate or, what is more basic, that those nations' authorities will even allow these systems to be hooked to the public telephone network.

Standards are Not Universal

Although the International Electrotechnical Commission (IEC) has established comprehensive guidelines for areas such as operator safety, there is no one widely accepted standard for safety protection for modems, CBF boards, and other telephone apparatus. Model rules may be drawn by international bodies and be agreed to in principle by the various countries, but they have not been implemented wholesale at the various national levels.

Thus, safety tests for CBF boards differ from country to country. The tests are carried out differently, the voltage requirements are different, the way the equipment is tested for overloads is different, the number of seconds the device under test is supposed to withstand the breakdown is different, and so it goes.

While coping with differences in telephone networks (such as one exchange needing pulse dialing with a 60/40 make/break ratio, while another requires 67/33) is not a major problem, other requirements can become quite complicated. Again, safety regulations are a good example. Although it would appear lightning strikes telephone poles in the same way in Finland, France and Germany, and cars knock down power lines on to telephone wires in the same way in Italy as they do in England, rules differ.

As with the United States, experiences with these situations have been unique for each country and the laws and regulations that have been established as a result reflect this. Testing procedures differ as well. Although European Economic Community (EEC) rules now theoretically apply equally across Common Market nations, standardization is yet to take place. It will be some time before it is possible to get a certificate out of a testing lab anywhere in Europe and have it cover equipment throughout the EEC. The general adoption of uniform EEC standards, especially for analog equipment, still appears far in the future. A homologation granted in the UK will not be good in Spain any time soon.

It is not surprising then that just collecting specifications needed for compliance can be a major, very expensive, undertaking. A veritable profession has developed around this need, and consulting firms specializing in the gathering of this information for companies seeking homologations have appeared all over the world, offering expertise of varying value and depth.

The Homologation Process

On the average, collecting information on the various specifications that must be met to receive homologation can range anywhere from 30 days to six months. This process not only involves gathering the information, but also having a native sponsor, a local address, contracts authorizing the people who are going to sell the product to act as your agents in the country, and other issues that vary from one national jurisdiction to another. None of this can be worked out from the United States. A local representation is a must. Even if all specifications are obtained and a prototype that follows them scrupulously, is then produced, by the time the product begins going through the approval process itself, it is not unusual to discover that in the interim there has been one or more bulletins invalidating or adding some of the parameters used to build it.

Technology does not stand still, and specifications and safety regulations evolve continuously to keep pace with these developments. This not only occurs in hardware and technical situations, but also in the more complex areas of protocols. The level of finished product that a country requires before certification is granted is also a variable. In some countries an absolute production-level product must be presented and no substitutes are accepted.

Other countries are satisfied with a one-of-a-kind prototype and will grant approvals based on that. Sometimes protocol testing is part of the approval process sometimes it is not. The situation has been encountered where the approval agency is solely interested in the telephone interface power levels and safety of the CBF board, and not in its actual operation as a fax device, or in its capability to communicate with other machines. Often, part of considering a product as being complete for the approval process is meeting the requirement for a full translation of all material and manuals connected with the product, including software screens. Generally, however, it is only required that user documentation be translated.

On the average, it takes six months to produce a prototype ready for preliminary testing at the target country. At this time, a contract is entered into with an approved testing laboratory in the target country that (just like its FCC-licensed United States counterparts), is authorized by the pertinent ministries to issue compliance reports. The CBF board prototypes are shipped, and the approval process then begins in earnest. The first result is a very detailed preliminary report outlining where the product must be improved or why it did not pass.

Once the preliminary report has been received and studied, it typically takes six to eight weeks of additional work to implement changes, tune up circuits, re-parameterize the software, deal with all the issues that were missed, and come back with another product iteration. A second contract is signed with the testing laboratory, and two to three more weeks are required to put the product through its paces again. Everything is retested: software, hardware, and circuit design. If the necessary changes will be reflected in the manuals, new translations of these must also be provided.

During this testing process it is worthwhile for the company seeking homologation to have at least one of its engineers present at the official laboratory, to ensure that the installation and use of the product are within specifications. Since the laboratory technicians do not have the time to learn the product, it is advantageous to have in place someone knowledgeable to assist with the installation, set things up, and then withdraw into the background ready to reappear if assistance is needed or further questions need answering. While the engineer's presence does not affect how the laboratory measures the product, it helps the process along by ensuring it is done correctly.

Finally, if nothing further is found wanting, the laboratory issues the compliance report. No compliance report is issued until everything even remotely connected with the product that will be imported complies with the requirements of all agencies: electrical, safety, communications, public telephones, user manual translations, etc. Getting a compliance report does not mean the product is homologated; only that it meets the requirements for homologation. The report now goes to the proper ministry (or ministries), to fulfill the legalities

219

of filing and posting. Six to eight additional weeks may pass from the time products are fully technically compliant and the report has been sent, before all the various ministries issue their own compliance reports to the PTT. Then it, in turn, produces the certification of compliance. It is now that the coveted approval number and import permit are granted.

Each complete homologation process can average from six months to a year, and cost anywhere from \$50,000 to \$100,000. In the end, it is anticlimactic to realize that after all this effort and expense you are authorized to sell in only that one country.

Why Do So Few Pursue Homologation?

It is logical to wonder whether, if homologation is of such value to products like CBF boards, why all companies do not pursue it with determination.

If a company is to seriously attempt wide scale homologation, it will soon learn that there is more to it than just determining what a particular country wants and then tinker it into the product as a kind of design afterthought. In our case, we made the decision when we first started that our software had to allow, among other things, the parameterizing of timings. The hardware is designed from the beginning enable circuits to accept modular components.
This is key to success in homologations. Our knowledge of the PTTs, acquired during those early attempts to get international approvals taught us the differences that must be watched for to successfully procure homologation. We realized that these differences lie principally in the signaling area what goes on to the telephone network in the timings of that signaling, and in the development of cadence monitors for it.

A CBF company expecting to sell its products internationally cannot get by with having one software that does do all it must to be fully parameterizable it for ringer differences, for how long the boards have to signal the specific telephone network. Every analog and digital circuit having anything to do with telephony must be fully programmable at the software level.

When someone looks at a CBF board designed with the international market in mind, one of the first things that is noticed, particularly in the DAA area, is empty slots, available for different modules. For a board to be sufficiently programmable it must allow the easy insertion of components for different countries, the adjustment of ring levels, and other parameters. This setup allows the unique country codes defined by the ITU-T to be used as a single trigger. This, as well as other capabilities guarantee customers that when they enter that number into the software, it will load the correct table for the country. This, in turn, programs the CBF card for correct operation with that particular nation's phone lines, as well as for meeting the various legal specifications that are required by each government. Developers using products designed with this kind of foresight are not burdened with these considerations it is all on the board and software, ready to respond transparently. There is no need to worry about things like unmatched signal levels, ringer differences, violation of access rules, and disruption of service, as can be the case with "grey market" products; that is, unapproved boards manufactured elsewhere that are brought into the country and used without knowledge or permission of the authorities.

Homologation and Users

Homologation directly affects a system integrator's ability to market products beyond a country's borders. To afford developers the widest global opportunity available, however, a manufacturer must not only be able to guarantee that the product will communicate internationally with any fax machine anywhere on the planet, but that it will give the added value of meeting all of a specific country's official technical requirements. For example, in France if a fax call is attempted to a phone number that is not that of a fax machine, by law the caller may not try again before a specified amount of time. To comply with the law, the system must automatically prevent the user from trying to send to that number again before the requisite amount of time established and required by the French PTT has passed. Other, or grey market, CBF boards continue retrying. The result is that the system integrators and businesses using them find themselves in violation of the law of the republic. This situation can result in stiff fines and even prevent a system integrator from ever importing into that country again. The user may be prohibited from utilizing the equipment again or, in some cases, the equipment itself may be confiscated.

How these often subtle different PTT requirements result in the handling of the transaction should be well and clearly outlined in the CBF boards basic documentation. These are all factors of primary importance. The product should insulate developers and users everywhere from worrying about these matters. The interface to the fax subsystem should bridge all the pitfalls involved in homologation and protocol questions, and do it automatically and transparently. Ideally, customers who develop systems on a manufacturer's Japanese CBF board, for example, should be able to market their products in the United States or Iceland knowing that the working of their software need not change and that all national requirements–technical and regulatory–are being fully complied with.

Other benefits of homologation, particularly for multinational corporate customers may not be as immediately obvious. By standardizing on systems using products that have a wide range of homologations, corporate leaders can make centralized MIS decisions. This enables them to cut costs secure in the knowledge that the deployment of these systems in the countries they operate in will not be obstructed due to technical or regulatory shortcomings. Obviously, no one is homologated in every country on the globe; however, consideration should be given to whether a specific product is present in the world's largest markets, where most corporations have significant offices and fax distribution needs.

Section IX

Purchasing a Turnkey Solution

**The SpectraFAX Corp
Special Request System**

Purchasing a Turnkey Solution

The SpectraFAX Corp
Special Request System

Introduction

So you've decided that the direct and focused delivery offered by Fax-On-Demand is the best way to deliver your message. You still have another big decision: to build or to buy? Engineering and building your own Fax-On-Demand system isn't an option for everyone. And even for companies that can afford to do so, it isn't always the smartest way to proceed. Sometimes it's better to purchase a turnkey system.

This is particularly true if your company does not have the on-staff expertise—programmers, electronics engineers and telephony specialists — all disciplines required to design a reliable, robust, system with all the functionality you need. But availability of personnel isn't the only consideration. Development time is another. You may be faced with a narrow window of opportunity, and your staff just can't be diverted from their primary mission. And even if you do have the staff available now, what about the future? Ask yourself if you are ready to commit resources to the ongoing system maintenance and enhancement needed to keep a sharp competitive edge as the demands of your work environment and technology change.

If you are limited by a shortage of on-staff expertise or time, right now and on an ongoing basis, purchasing a turnkey Fax-On-Demand system can be a smart move — when you go about it in the right way.

Getting Started

Once you've assembled a list of potential vendors, start by checking their Fax-On-Demand demo lines. Besides gathering basic information on the turnkey solutions that are available, you'll be getting a feel for how each vendor's system will be able to represent your company and your products and services.

Like any purchasing decision where a significant investment is involved, a few common sense rules apply. Of course, price is a consideration. (Systems can run anywhere from $5,000 to $150,000 and up.) But price shouldn't be your only criterion. Compare the full range of product features. Consider, too, the likelihood that the vendor and his product will be able to support you over the life of your fax application.

Above all, evaluate vendor reputation and expertise. Make sure the vendor you select has successfully handled top name customers running high-profile, high-value, mission critical applications. Then take the time to check their references.

Doing Your Homework

Although most knowledgeable vendors will be able to help you refine your estimates, you'll want to have some idea of what size Fax-On-Demandsystem you need for your application. There are two measures of system size: the number of voice/fax ports (sometimes called "channels") provided by the system, and the amount of hard disk storage space available to hold the materials you wish to make available to your callers.

Port Sizing

How many ports you'll want to include in your system is determined by the number of calls you anticipate during peak calling hours, and the average connect time for each call. Table 1 shows the number of voice/fax ports needed to handle various numbers of calls during the peak usage hour, assuming various connect times per call.

Avg Total Connect Time per Call (in secs) During Peak Usage Hour				
60	120	180	240	300
Number of Ports				
4	60	30	20	15	12
6	120	60	40	30	24
8	214	107	71	53	42
16	561	280	187	140	112
24	940	470	313	235	188
32	1340	670	446	335	268
48	2150	1075	716	537	430

Table 1. Port Requirements as a Function of Total Connect Time

To use the table, find the average connect time you anticipate for calls on your system in the top horizontal row (Average Total Connect Time per Call). For most applications, the average call lasts about 120 seconds. This includes about 60 seconds for voice prompts and subscriber choices, and sufficient time to send three fax pages to the caller (an average fax page takes 15 to 18 seconds to transmit). If your experience suggests otherwise, move to the left (to a shorter connect time) or to the right (to a longer connect time) on the "Average Connect Time" row.

Next, look down the column beneath your average connect time to find the value nearest the number of Fax-On-Demand calls you anticipate during the busiest hour of the day. Finally, move left to read the number of ports you will need. For example, with an average connect time of 120 seconds, and an anticipated 470 calls during the peak usage hour, you will need 24 voice/fax ports. Note that this table assumes a 3% blocking rate—three out of every 100 peak-usage-hour calls will receive a busy signal and will not get through. You can add ports to reduce the chance that your callers will get a busy signal.

Fax-On-Demand systems can be either stand-alone or networked. Stand-alone systems will generally accommodate up to 8 ports. If you need more than 8 ports, you should consider a networked system for best response and ease of administration. A networked system can provide hundreds of ports, divided among multiple nodes. A networked system is an economical choice if you anticipate rapid growth. Start out with two or three nodes and expand the number as more ports are needed. If larger call-handling volumes become necessary, individual networks can be bridged together.

Estimating Your Document Storage Capacity

Your next step is to estimate how much hard disk storage space your system will require to hold faxable copies of all the documents you wish to make available to callers. Use Table 2 to estimate the amount of hard disk storage you will need, based on the number of pages you plan to make available.

Hard Disk Capacity	Page Size (in KB)			
	40	60	80	100
245 MB	5,600	3,700	2,800	2,200
560MB	13,500	9,000	6,700	5,400
1GB	24,500	16,300	5,400	9,800

Table 2. Hard Disk Size as a Function of Pages To Be Stored

Enter the table at the "Page Size" value that is closest to the typical page you will be offering. On the average, a single-page fax document takes up 60KB of storage. If your documents incorporate detailed graphics, move to the right, to a denser page. If your documents are mostly text, move to the left. Once you've settled on an average page size, move down the column to the number closest to the total number of pages you anticipate needing to store. Be sure to allow for growth. When estimating disk storage needs, it's always wise to err on the generous side. Finally, move to the left to read the hard disk capacity you will need.

For example, assume you have 7,000 pages of documents to make available, most of which are an average mix of text and graphics. Move down the 60KB "Page Size" column to 9,000. Then move left to the "Hard Disk Capacity" column. You will need a 560MB disk drive to comfortably handle your storage needs.

Now, during planning, is a good time to consider building redundancy into your system, especially if it will be playing a mission critical role. Many Fax-On-Demand providers mirror their hard drives and controllers so that if one drive should fail, the other can take over immediately. At the very least, make plans for regular backups.

Hardware Considerations

The port-sizing guidelines outlined earlier assume that each port can handle both voice and fax — and with good reason. With separate voice ports and fax ports, idle voice ports can't pick up the slack when the demand for fax ports exceeds supply, and vice versa. But by combining inbound and outbound capabilities in each port, you can, under extreme conditions, effectively double the capacity of your system.

The capabilities of the fax and voice boards built into a Fax-On-Demandsystem directly impact performance and operational costs. Hardware and associated firmware determine system characteristics such as:

- **Transmission Speed**: Most fax machines can handle a maximum transmission speed of 14,400 bps. Your system should be able to do the same.

- **Fax Compression:** The compression techniques employed by a Fax-On-Demandsystem determine the size of the fax file, for storage and transmission purposes. During transmission, higher compression means faster transmission. For storage, higher compression means smaller files and better use of disk storage.

- **Call Progress Monitoring**: With call progress monitoring, a Fax-On-Demandsystem can "listen" for tones during outbound calls, and, depending on what it "hears," take appropriate action for greater efficiency and cost savings. Some of the situations call progress monitoring can identify are whether the called number is busy, is ringing, or is answered by fax or voice.

Look for quality in the design of the hardware components that make up the Fax-On-Demand system. The well-known manufacturers will be using state-of-the-art board manufacturing techniques, such as integrated voice-fax cards and surface mount technology, to reduce chances of hardware failure and increase overall reliability and robustness of the system. In the long run, a quality product will mean fewer operational problems and less downtime.

Be sure to review the terms of the vendor's warranty. A vendor's willingness to stand behind the equipment he sells is a good indication of its quality.

Software Considerations

The features and quality of the software are truly what distinguishes one system from the next. When it comes to evaluating the software for different Fax-On-Demand systems, your time can probably be best spent looking at a special piece of software called the script language. It is the script language that lets you define the application your system is going to perform and how it will interact with your callers. It is the script language that allows you to control and customize the functions of your system. Some of the characteristics of a good script language are:

- **Flexibility.** Will the script gracefully accommodate a wide range of calling situations, for example, one- and two-call faxing, calls with large numbers of caller errors, and callers who know beforehand what menu selections they want?

- **Functionality.** Does the script language let you do all the things you need to do right now as well as providing room to grow as you find new uses for the system?

- **User friendliness.** Is the script language intuitive and easy to use? Will extensive training be needed before your staff can modify and adapt applications and write new ones?

The checklist at the end of this article can give you some idea of the features you should be looking for in a script language.

You'll want to hit the ground running with your new Fax-On-Demandsystem. As part of the cost of the system, insist that the vendor provide a fully tested custom first application. Look for a vendor who can provide timely script-writing assistance in the future, should you need it.

233

One final bit of software-related advice. Find out whether the vendor controls their own code. If the vendor is only a "middleman," you may run into a software problem and find yourself at the mercy of a third party.

Connectivity

Obviously, the Fax-On-Demand system you select must be compatible with your phone system. But looking beyond the basic requirements, be sure the system you choose is designed to take advantage of the capabilities and features of state-of-the-art telecommunications services — for example, ANI (automatic number identification), in-band signaling, DNIS (dialed number identification service) and DID (direct inward dialing). Check with potential vendors about compatibility of their systems with your telecommunications environment, particularly your PBX and voice mail system. If you have, or expect to set up, international operations, investigate the vendor's experience connecting with international telephone systems, too.

In addition to telephony connectivity, the system you select should offer connectivity to your existing computing environment, whether it be built around an IBM mainframe, DEC or UNIX minicomputers, or LAN servers. The ways you can exploit the connectivity between your system and your company's host computers are practically unlimited: order processing, data collection, and customized faxes, to name a few. Don't shut yourself off from the benefits of connectivity by selecting a system that is either incompatible or not easily adapted to your computing environment. Keep in mind that effecting smooth interaction between Fax-On-Demand systems and host computers is not a trivial accomplishment, and not all vendors can offer support in this area. The vendor who can link your Fax-On-Demand system to your host for real-time delivery of information and database update has a definite advantage.

Installation, Training, and Support

Depending on the Fax-On-Demand system you need, installation can be fairly simple or exceedingly complex. A four-port stand-alone system, for instance, may require only that you plug in to an electrical outlet and four phone jacks. On the other hand, a large, 48-port networked system, integrated with your PBX and host computer can present complex installation challenges. Your vendor should provide professional installation for any but the most basic systems.

As with any new technology, once you make your purchase, a little hand-holding is in order. Check your vendor's training and support policies. As part of the purchase price, make sure you get adequate basic training so that your people can handle routine maintenance and make use of all the system features you are paying for. Consider advanced training courses, too, especially if you plan to maintain the system yourself.

Inevitably, you will have questions as your system gets put through its paces. Complete and well-written technical documentation is a must. In addition, your vendor should provide accessible and knowledgeable telephone hotline support. Vendors who are truly committed to customer service will also offer remote logon capabilities for on-line problem diagnosis, real-time "fixes," and software upgrades.

Administration and Maintenance

High administrative overhead can quickly cancel out any advantages in performance and convenience. The behind-the-scenes effort needed to keep a Fax-On-Demand system operational should be minimal. Although virtually no routine hardware maintenance should be required, the system should incorporate features and utilities to speed and simplify administrative tasks. Among these tasks are configuring the Fax-On-Demand system to work with your phone system, collection and analysis of system activity data (including reports), and creating and managing faxable documents.

To make the most of your system, you'll want to know things like which documents are in greatest demand, who your most frequent callers are, whether your system is nearing capacity, which applications draw the most traffic, and why a specific call failed. These and other kinds of information are made available by a system's activity logging and reporting capabilities. Any solution you seriously consider should have good data collection features, along with system activity and system usage reporting tools. Moreover, make sure you can export log data to your database program for more extensive custom analysis.

Fax-On-Demand systems require that your document offerings be in specific file formats. ASCII text format and either PCX or TIFF graphics formats are most commonly required. Creating the initial library of documents may involve scanning hard copy or using conversion programs on existing files. But preparing documents for faxing will be an ongoing effort, and you'll be ahead of the game if your Fax-On-Demandsystem can convert file formats in real time. Also look at what built-in tools a system offers for checking correctness of graphics file formats, for combining single page documents into multipage files for efficient transmission, and for batch conversions.

Phone configuration includes telling the Fax-On-Demand system what extension is assigned to each port, which application callers will reach when they dial a specific extension, and how to distinguish and dial local and long distance calls. You will save time if the configuration program is straightforward and easy to understand.

Keeping Competitive

You can expect the system you select to be around for a while, so it's important that your system be scalable. Once your target audience sees how convenient Fax-On-Demand is, it's almost a sure thing that the number of callers will increase over your initial estimates.

You and others in your company are certain to find new applications for fax on demand. The evolving mobility of the workforce points to a more central role for the "anywhere, anytime" convenience of Fax-On-Demand information retrieval. And the immediacy of fax fits right in with the corporate trend toward reducing each and every time wasting factor. Experience shows that as other departments see how effective Fax-On-Demand can be at cutting costs, they're quick to find their own applications for the system as well. Your choice of system needs to be able to grow to accommodate more ports and more storage. As traffic increases, make sure you're not caught having to rip out the old system to install something completely new.

Besides being able to expand to handle increased call volumes and new Fax-On-Demand applications, find out if the same hardware and software platform that provides Fax-On-Demand can support other enhanced fax products. In particular, can you investment be leveraged to include these other offerings:

- Fax Broadcast - A cost effective way to send your message to a large number of people.

- Fax Mail - Store and forward message service with remote retrieval capabilities.

- Enterprise-wide fax routing.

- LANFAX - With the rapid proliferation of corporate LANs, you'll want to be able to consider inbound and outbound fax capabilities for workstations using a shared fax server.

- Fax-as-EDI (electronic document interchange) - Fax server integration with computerized forms and workflow.

- Automated data entry via fax and OCR (optical character recognition) or the more advanced ICR (image character recognition).

- Fax server-based publishing and telemarketing.

- Fax servers as intelligent gateways to image-document processing systems.

237

And the list is growing. Your vendor should be committed to ongoing development to help you keep competitive. As telecommunications and information technology evolves, you want to be able to take advantage of the benefits offered.

Caveat Emptor

Your choice of a turnkey Fax-On-Demand solution will have an impact on your business for years to come. With planning and forethought, it can be an integral part of your strategic line-of-business applications. Because of this potential, it pays to approach the selection task analytically and with a clear understanding of the issues. The following checklist, though far from exhaustive, will help make sure you consider some of the key issues in making your selection.

Turnkey Fax-On-Demand Checklist

Vendor Qualifications
Experience with high-profile mission-critical applications
Outstanding references (Be sure to verify them!)
Liberal warranty terms (hardware, software, labor, travel)
Ongoing product development program

Script-Enabled Features
Flexible script language
Fully-tested first application
One-call and two-call fax delivery
Document choice from faxed index or voice prompts
Scheduled future fax delivery
Calling restrictions (by area code, telephone number, country, etc.)
Passwords to control access to applications
Typeahead (to reduce phone connection time)
Ability to limit number of documents per call
Detect and prevent duplicate faxes in a single call
Allow transfer to live operator
Caller error handling
Credit card validation
Voice confirmation of selections
Math functions
Evaluation of strings
Remote file input for re-transmission
Cover pages, logos, signatures
FROM and TO information on faxes
Alpha key entry
Script-writing assistance available

Performance Feature
Combined voice/fax ports
Call progress monitoring
High-speed file transmission (9,600 bps, 14,400 bps, or better)
Fax compression
Fax resolution: Standard, Fine, Extra Fine
Real-time document format conversion

Administrative Features
Voice recording and editing utilities
Phone configuration tools

System activity logging
System activity and system usage reports

Scalability and Expandability
Document storage capacity
Number of fax/voice ports
Support for additional enhanced-fax products

Connectivity
Connectivity to your host computer environment
Connectivity to your PBX and telephony environments
International telephone system connectivity
Connectivity to your electronic mail system

Training and Support
Good documentation
Hotline phone support
Vendor control of code
Remote logon for system diagnostics
On-site basic training
Advanced training

Price

Section X

The Future of Fax

The Future of Fax

Group 4

In the swift evolution of fax technology, the Group 4 standard is viewed by some as the next significant step. Proposed in 1987 by ITU-T standard-makers, it was to allow the transmission of fax over non-analog circuits using X.25 packet switching networks as well as others. Group 4 is expected to work with ISDN, switched data networks, or dedicated digital circuits. It is a layered standard, laid out in terms of the Open Systems Interconnection Reference Model, and is expected to provide high-speed transmission, outstanding error control, and features such as gray-scale and color imaging.

Group 4, however, is a standard written before a real machine was made and only a group of experts well-versed in its arcane ins-and-outs know how to implement it. Currently, there is no way to use it over ordinary voice grade phone lines, and it will not interoperate with Group 3, 2, or 1 machines. At present, there is no available modulation standard or scheme for Group 4. Absent, also, is a protocol for initial handshake. Applications for Group 4 seem rather limited, particularly in view of Group 3 technologies continued evolution.

Group 3 is now expected to provide many of the features once promised only by Group 4, such as tighter compression, gray scale, color, and faster transmission. Work has been concluded by the ITU-T on a new, faster modem standard, V.17, which gives Group 3 the capability to operate in the 12,000 to 14,400 bps range. Many industry experts are dubious about Group 4's future, arguing that the continued evolution of the T.30 protocol and Group 3's capability to deliver speed, error control, and higher resolution the very features that made Group 4 attractive has rendered Group 4 outmoded even before its implementation in the real world.

Primary objections to Group 4 are its inability to communicate with Group 3 machines and work over ordinary telephone lines. Before consideration can be given to any radical alterations in fax standards, the first thing that must be dealt with is the existence of some 25 million Group 3 machines installed all over the world today. This is a considerable installed base, particularly when one considers these machines are not upgradable nor can they be changed in any significant way. What is more, they run satisfactorily, and users show no intention to discard them for several years to come.

Given that there is such an enormous installed base of unalterable machines, what is going to happen? They cannot be upgraded by adding new keyboard features or new LCD displays, they will not accept new software, there is nowhere one can send in a few dollars and get a new control panel for them. And, like Everest, they are there. Even if the protocol committees were to come out with faster modem definitions, higher resolution pages, etc., a fact of life is that these machines are going to be around well into the 21st century.

The only way users might be led to discard this installed base in any significant way would be if there were some truly fundamental economic reason to do so. This happened to the majority of Group 2 fax machines which were disposed of because transmission time for Group 3 was a third or less than for Group 2. There was a significant telephone cost difference in going from a three-minute to a one-minute fax. The more knowledgeable experts do not believe, however, that this will happen with Group 3.

Once the fact that there exists a large infrastructure of installed machines that is not going to change is accepted, and that any software innovation and fax system planning has to take place taking that reality into consideration, it becomes easier to plan what uses to put the technology to, as well as what sorts of services to offer with it. Thus, services like fax on demand and others have come into being around that reality. Hardly a day goes by that new ways to use fax do not come into existence. These are very innovative and useful applications, but do not require changes in Group 3 technology.

Group 3 Evolves

That being stated, it must be also pointed out that Group 3 technology is not static, but is also evolving. There are extensions being added to it. These involve agreements by the standard-making bodies from various countries and administrations and, ultimately the worlds manufacturers, for additions to the Group 3 protocol set. Group 3 fax machines now boast faster, 14.4 kbps modems. It is estimated that now perhaps two to four percent of the installed base consists of fax machines with the higher speed modems, and this is increasing. Also coming are new resolution capabilities. It is expected that soon, 300 dpi resolution will be approved by the standard-making bodies. Some of these machines may make their appearance as early as next year.

As the technology continues evolving, new capabilities are added to Group 3: there are new routing codes, binary file transfer capabilities, and superfine resolution, just to mention a few. Although market inertia is considerable, some of these new features may be of particular importance to some users, who will invest in machines offering them, slowly obsoleting existing machines. Higher speed modems and error-correction were features first introduced not too many years ago, which are now well-entrenched in the marketplace.

Where is CBF Going?

It would appear from all this that computer-based fax by being, part of a computer and therefore more amenable to enhancements and software upgrades would find itself in a privileged position. This is the case, up to a point. The fact remains that an estimate of all LAN fax servers, and even all the PC fax boards, sold barely reaches the half-million mark. It is projected that by the end of the current year there will be about a million of various versions of CBF installed. Although this is by no means a small amount, and market acceptance is growing literally by leaps and bounds, the fact remains that anyone wanting to use the capabilities of these computer/fax hybrids must still deal with an overwhelming installed base of external standalone units that cannot change. So, while software for laptops or desktops can add features to CBF, unless these features can find their way to the rest of the industry, into the installed base, they will evolve very slowly.

The Digitalization of Fax

It is very clear that there is a trend well-underway leading to digital fax by the end of the century. Everything seems to be working toward it the economics of the technology, as well as the integration of fax with the computer/communications community and other networks and circuits available for the transmission of digital data. To take advantage of all this, fax must become digital. The only question is whether this is going to require yet another new standard, or whether Group 3 will gradually evolve into a form of digital standard. The extensions under proposal suggest that Group 3 itself may migrate and become the all-digital standard.

Probably the biggest growth area for fax today is in the area of its digitalization. There are several proposals under consideration by the various standard-making bodies, which provide for the transmission of fax over non-analog circuits. Group 4 was the first one of these. There are Japanese fax machines sending Group 4 fax over ISDN B (Bearer) circuits at 64 kbps. However, in practice, it is only in Japan that any significant population of these fax machines is found, and even there it is very small, in the tens of thousands.

The reason is that Group 4 requires digital circuits that are not commonly available. Also, these fax machines have 400-dot print scanners and sell for about $10,000 or $12,000, a price most businesses find difficult to justify, particularly when the ongoing additions and expansions taking place within the Group 3 protocol continue making Group 4 seem less desirable.

While Group 4 machines are intrinsically faster, the speed difference between a 14.4 kbps and 64 kbps message may not be all that critical. While it is true that instead of taking 30 seconds to send a page a Group 4 fax can do it in ten, it is still necessary to wait 20 seconds for call setup procedures; Group 4 is a very complex protocol, machines must exchange some 20 messages before they begin transmitting and receiving. An industry observer compared Group 4 to the Concorde: A three-hour flight to Europe does not make that much difference when you spend four hours waiting in the airport. And, of course, there is that $12,000 price tag and the fact Group 4 machines do not operate over ordinary telephone lines and have no compatibility with the great infrastructure of installed Group 3 machines.

There are other digital standards, such as Group 3 bis and Group 3/D, which keep the standard Group 3 protocols and resolutions, provide fast handshaking, and make all digital, full-duplex transmission possible. These standards are also ISDN-compatible. The X.38 and X.39 standards offer another way of digitizing Group 3 fax, working with an X.25 transport layer.

The Merging of Fax and X.400

X.400 is a store-and-forward electronic messaging protocol defining a framework for distributing data from one network to several others. It allows end-users and application processes to send and receive messages which it transfers in a store-and-forward manner. An X.400 message consists of a message envelope and message content.

The message envelope carries addressing, routing, and all control certification information. The message content part involves methods of encoding simple ASCII messages, as well as more complex data. It may contain fax, graphics, text, voice, or binary data structures. X.400 is being increasingly viewed as a delivery platform for a variety of services, including E-mail, electronic data interchange (EDI), and others.

Industry observers agree X.400 fax will play a major role in the next generation of facsimile machines. They look at a combination of fax and digital fax with the X.400 E-mail standard. In such a setup, a Group 3 message would be packaged as part of an X.400 E-mail message and sent over the X.400 hierarchy of circuits. There are presently over 100 international carriers that can exchange this type of traffic. X.400-based fax store-and-forward networks are already in operation. The user would transmit a fax locally to a store-and-forward switch that would capture, encapsulate, and send it over high-speed circuits to be recreated later, as a fax and delivered.

The concept of sending fax raster images as part of an X.400 E-mail message makes sense, because X.400 already exists as a world standard. It is now in its third division and no further standards have to be implemented to put in fax raster into X.400 traffic. X.400 fax traffic would parallel E-mail traffic generated by corporations. It is not unlikely that in the near future, traveling side-by-side, there would be an E-mail message followed by fax, and a combined fax and text message, then by a voicemail mail message, etc., all of these intermixed in one circuit. Compressed TV, and animation are not too far off, either. In short, X.400 will become the ideal pipeline for multimedia communications.

250

The problem with X.400 today is that it is principally used as a backbone to connect disparate email systems. There are few good user interface packages available, addressing is extremely complicated, as are the options. The tendency has been to install LAN mail systems that use X.400 if needed, and X.400 has not gained the role it deserves.

As an industry observer put it, It is just the Esperanto used to get all these email systems communicating with each other. Another problem with X.400 today is that it is primarily a store-and-forward system. One of the things people like most about fax is its realtime immediacy. After receiving a transmission confirmation, the user walks away from a fax machine with 99.99 percent certainty that a piece of paper is in the persons machine at the other end, and that if he picks up the telephone and calls to discuss the fax, it will be there.

This is not always the case with X.400. Although the user may get a local transmission confirmation, it may be another 20 minutes before the message works itself across the various hierarchies of X.400-land and arrives at the recipients mailbox. It may find itself at the end of a queue waiting for channels to clear, or pause behind a long file transmission. The existing X.400 standard does not have the immediacy that fax users have grown accustomed to expect. This is not an insurmountable problem in many applications such as the broadcast of faxes overseas, however, and X.400 will survive and evolve as a real commercial, end-user standard. Eventually, there will be a large EDI population of users. It is a superb backbone for E-mail systems, and recognition of its capabilities to handle fax and raster traffic is slowly taking place.

Very little color facsimile equipment is available at present. Among the reasons for this are the complexity and cost associated with color scanning and printing, as well as a long transmission time. Another obstacle to color fax is a lack of standardization for areas such as color components, transmission sequence, bilevel vs. continuous tone, and encoding algorithms.

Widespread use of color fax will probably take place within the X.400 framework, because of this architectures evolution into a multimedia

E-mail vehicle. X.400 is, for all intents and purposes, evolving into the international standard for multimedia communications.

Remaining Obstacles

Forward-looking corporations have already come to the realization that in order for fax to achieve a flawless integration with telephone and computer technologies, one of the things that is going to have to take place is for both these communities to stop looking down at fax as an orphan of the other. The merger of these technologies has to take place. The computer world with Windows and OS/2 and the Macintosh has gone the way of raster, bit maps, and graphical images. Fax and fax machines can no longer be considered as a separate communication service with a separate network of its own. Image traffic from Windows workstations on computer networks and raster imagery generated by fax machines must merge. It is no longer practical for a company to have a telecom department that deals with fax problems, separate from its MIS department which handles computer and email traffic. These areas must converge: they are holding back people and the technology. There can no longer be disparate networks.

As requirements increase, and faxes become digitized, a form of multimedia fax will make an appearance, providing high resolution, color, working with email, and offering the security of encryption. All the pieces are available, all that is needed is for consumer demand to reach critical mass.

New CBF Markets

As fax technology continues developing and finding new applications, new markets appear to fill the demands of business, government, and personal users. While by no means all-encompassing, below are what is generally considered as the major growth and application areas for computer-based fax (CBF).

Network Fax Servers

As more companies rely on fax communications to get their work done, network fax servers are providing them with the means with which to control and optimize their fax communication. With a network fax server, any business that regularly sends and receives faxes can increase productivity while saving time and money.

Network fax servers make it possible to send faxes right from the users workstation. Many fax servers even let the user fax documents right from the applications that created them. The user simply indicates the fax number, and the fax server does the rest, leaving the person free to resume work.

Incoming faxes can be directed back to individual workstations as well. Different fax server products offer different ways of handling incoming faxes: a fax can be received by an operator who then forwards it to the individual workstation, a unique fax extension can be assigned to each workstation for TouchTone routing, or the individual direct fax telephone line can be accessed through a Direct Inward Dial (DID) line.

EMail Fax Gateways

As companies look to bring CBF to their computer network, many find it easier to implement and train users on a system when it is simply an extension of their electronic mail. With an email fax gateway a business can expand its mail system by adding fax as an additional messaging capability. This extends the convenience of email communications beyond the networks boundaries.

With an email fax gateway, all users can send and receive faxes from within the email package. Because the fax messaging is integrated directly into the system, fax users have virtually no new software to learn. With one place to turn to for all messaging needs, users have all the features that they have come to depend on in email, such as workgroup distribution lists, the capability to attach graphics files to text messages, support for different-user operating systems on the network, and instant notification of received messages. With an email fax gateway, the same message can be sent to both mail users and fax recipients.

Mini/Mainframe Fax Servers

When it comes to CBF communications, each business has different needs. Although many companies generate and store information on mainframes or minicomputers, most CBF hardware is PC-based. With third-party applications running on PC front ends, CBF can meet the communications needs of these customers, bringing the benefits of CBF to the mini/mainframe world.

Many mini/mainframe fax servers work in the same way as LAN-fax servers: users can send and receive faxes from their terminals, saving both time and money while improving the quality of their fax transmissions.

In addition to providing fax services to user terminals, mini/mainframe fax servers have carved out a niche for themselves through direct integration into specific applications. By being on the same computer system as an organizations data, mini/mainframe servers are positioned to provide an EDI-like component for large, vertical applications such as accounting or purchase ordering, broadening the electronic reach of these programs to any recipient with a fax machine.

Network Fax Servers

As more organizations rely on fax communications to get their work done, network fax servers are providing them with a way to control and optimize their fax communication. With a network fax server, any business that regularly sends and receives faxes can increase productivity while saving time and money.

With a network fax server, users send faxes right from their network workstations. Fax servers let users fax documents from within the applications that created them. The user simply indicates the fax number, and the fax server does the rest, leaving the user free to continue working.

Incoming faxes can be directed to a workstation as well. Different fax server products offer different ways of handling incoming faxes, but most use one of the following: a fax can be received by an operator who then forwards it to the appropriate user workstation, users can have fax extensions assigned to their workstation for TouchTone routing, or users can have their own direct fax telephone number that is accessed through a Direct Inward Dial (DID) line.

Key Selling Points of LAN Server/Gateways

Cost Savings

The major advantage of LAN Server/Gateway implementation is the significant savings that can be achieved by organizations in three areas:

Labor savings: When compared to sending a fax via a fax machine, organizations can achieve significant savings in productivity gains from its employees as depicted in the Labor Savings Chart. The savings achieved simply from increased employee productivity delivers a return on investment in less than 90 days.

Delayed Sending Savings: An often overlooked savings that can be achieved by organizations is in the reduction of phone charges by taking advantage of the delayed send capabilities that many fax server and fax gateway software vendors have implemented. These savings can be significant for organizations that have fax traffic where differences in time zones does not warrant a fax be sent immediately during peak hours at peak costs.

Savings on Consumable Products: Savings on consumables can be achieved in both the sending and receiving process. Because users do not need to first print documents to be faxed, savings will be realized in the purchase of paper. On the receiving end, users can review faxes on their monitors and print only those faxes of which they need a hard copy. Additionally, users can simply route the electronic version to others, resulting in even less paper usage and reduced copy machine usage.

Connectivity Savings: The centralization of equipment for transmission and reception, coupled with automated transaction processing and handling, allows for the reduction in the number of phone lines required for a more efficient handling of fax communications.

Automation Benefits

In addition to centralizing equipment and reducing the number of phone lines needed for fax, LAN server/gateways provide additional benefits normally associated with automation:

Billbacks: Customers in the service industries and organizations whose accounting procedures bill charges back to departments can receive accurate billing reports with no additional data entry or paper work.

Transaction Statistics: Organizations can track, monitor, and adjust fax transactions to become more cost effective.

New Communications Capabilities: Organizations can improve overall communications with customers, vendors, and remote employees using fax broadcasts from their LANs for distributing information. The use of corporate phone books reduces the possibility of incorrectly entered fax numbers.

User Benefits

Not only does the overall organization benefit but employees also have much to gain with the implementation of LAN server/gateways. These gains are primarily in three areas:

Convenience: Unprecedented convenience of sending and receiving faxes is one of the most powerful user benefits. Users no longer have to print documents, stand in line at a fax machine, or be concerned about reaching busy fax numbers. Network fax is as simple as printing.

Quality: Because the document that is faxed is an original, the receiving document has not degraded due to the scanning process. And the ability to send laser-quality documents ensures the readability of the receiving fax.

Confidentiality: Sensitive material such as contracts, personnel records, and financial information can be sent and received without the fear of unauthorized viewing.

Glossary

Why reinvent the wheel....

This Glossary is a compilation of three sources:

Newton's Telecom Dictionary, by Harry Newton
Published by Telecom Library Inc.
212-691-8215 or 800-LIBRARY
Harry's dictionary is the first book in any CTI/telecom library.

PC-Based Voice Processing, by Bob Edgar
President, Parity Software Development Corporation
Published by Newton's Flatiron Publishing Inc.
212-691-8215 or 800-LIBRARY
This was the first and continues to be the best book on Voice Processing.

Understanding Fax Technology
GammaTech Publication GTP1001
GammaLink/Dialogic Corporation
Sunnyvale, CA
408-744-1400

This section, as well as this book, could not have been completed without their assistance.

10 BASE 2
IEEE standard for baseband Ethernet at 10 Mbps over coaxial cable to a maximum distance of 185 meters. Also known as "Thin Ethernet"

10 Base-T
An IEEE standard for operating Ethernet LANs on twisted pair wiring that appears like telephone cabling. Sometimes old cabling will not work.

A4
Basic Group 3 standard defined for the scanning and printing of a page 215 mm (8.5 in) wide. An A5 page is 151 mm (5.9 in) wide, and the A6 is 107 mm (4.2 in) wide.

ACTIVITY REPORT
Provides a record of transmission time, date, size of the file, recipients telephone number, transmission success or failure, the senders name, and other pertinent information. This is a valuable management tool to get an overview of a company's fax traffic and costs.

AEB
Analog Expansion Bus. Dialogic's name for the analog electrical connection between its network interface modules and its analog resource modules. This is an open technical specification, and individuals can make their own resource modules and/or network interface modules. See also PEB, which is the more modern digital PCM expansion bus.

ANALOG
Comes from the word "analogous," which means "similar to." In telephone transmission, the signal being transmitted — voice, video, or image — is "analogous" to the original signal. In other words, if you speak into a microphone and see your voice on an oscilloscope and you take the same voice as it is transmitted on the phone line and ran that signal into the oscilloscope, the two signals would look essentially the same. The only difference is that the electrically transmitted signal (the one over the phone line) is at a higher frequency. In correct English usage, "analog" is meaningless as a word by itself.

But in telecommunications, analog means telephone transmission and/or switching which is not digital. See ANALOG TRANSMISSION.

ANALOG EXPANSION BUS
See AEB.

ANALOG / DIGITAL CONVERTER
An A/D Converter. Pronounced: "A to D Converter." A device which converts an analog signal to a digital signal.

ANALOG TRANSMISSION
A way of sending signals — voice, video, data — in which the transmitted signal is analogous to the original signal. In other words, if you spoke into a microphone and saw your voice on an oscilloscope and you took the same voice as it was transmitted on the phone line and threw that signal onto the oscilloscope, the two signals would look essentially the same. The only difference would be that the electrically transmitted signal would be at a higher frequency.

ANI
Automatic Number Identification. A phone call arrives at your home or office. At the front of the phone call is a series of digits which tell you, the phone number of the phone calling you. These digits may arrive in analog or digital form. They may arrive as touchtone digits inside the phone call or in a digital form on the same circuit or on a separate circuit. You will need some equipment to decipher the digits AND do "something" with them. That "something" might be throwing them into a database and bringing your customer's record up on a screen in front of your telephone agent as he answers the call. "Good morning, Mr. Smith."

ANSI
American National Standards Institute. A non-government standard-setting organization which develops and publishes standards for use in the U.S.

API
Application Program Interface. A set of standard software interrupts, calls, and data formats that computer application programs use to

261

initiate contact with network services, mainframe communications programs, or other program-to-program communications. APIs typically make it easier for software developers to create the links an application needs to communicate with the operating system or network.

ASCII

American Standard Code for Information Interchange. A character set which gives a numerical value from 32 to 127 to commonly used letters, numbers and symbols. For example, an upper-case A is assigned the value 64. The IBM PC extended this standard to 255 characters which included symbols required in countries other than the USA such as accented letters. This extended character set is sometimes called the 8-bit ASCII character set to distinguish it from the original standard, which is then called the 7-bit character set since values from 0 to 127 can be represented using 7 bits, and can therefore be transmitted over modem lines using 7 rather than 8 data bits, for example. Computer files stored using these characters are called ASCII text, or sometimes just text files. These files almost always use one byte (an 8-bit binary unit which can take a value from 00000000 = decimal zero to 11111111 = decimal 255 when interpreted as a number) for each character. ASCII files are simpler than word processing files, which have complex codes embedded within them, and have the advantage that most programs can read and write in this format. However, there are only ASCII codes for the most rudimentary formatting information: tab, end of line and end of page markers. All other information such as font size, tab positions and so on is lost when a file is stored as ASCII.

ASCII CHARACTER SET

A character set consisting only of the characters included in the original 128-character ASCII standard. Every character handled by a computer has a number in the ASCII character set. Some computers use an additional 128 ASCII characters beyond the original 128, many of them dedicated to graphics. This is what is referred to by the term "extended character set."

ASYNCHRONOUS COMMUNICATION
A method of data communication in which the transmission of bits of data is not synchronized by a clock signal but is accomplished by sending the bits one after another, with a start bit and a stop bit to mark the beginning and end of the data unit. Two communicating devices must be set to the same speed, or "baud rate." Asynchronous communication normally is used for transmission speeds under 19,200 baud. Because of the lower communication speeds, normal telephone lines can be used for asynchronous communication.

ATM
Asynchronous Transfer Mode. Also known as "BISDN" and cell relay".

AUTOMATIC COVER LETTER
Feature that allows the user to automatically attach a cover letter to the document being faxed. This is especially convenient when sending material such as spreadsheets, for example.

AUTOMATIC REDIAL
Provides for the automatic redialing of a fax number in the event the receiving line is busy or an error occurred in faxing the document. Some products allow the user to specify the redial attempts and correct specific errors to be used when redialing.

AUTOMATIC ROUTING
Allows incoming faxed documents to be automatically routed to the addressed individual on a LAN or centralized system. Currently, technology provides several methods of accomplishing this using DID, OCR or DTMF techniques.

APPLICATION PROGRAMMING GENERATOR (API)
A program to generate actual programming code. An applications generator will let you produce software quickly, but it will not allow you the flexibility had you programmed it from scratch. Voice processing "application generators", despite the name, often do not generate programming code. Instead they are self-contained environments which allow a user to define and execute applications.

263

AUDIO MENU

Options spoken by a voice processing system. The user can choose what he wants done by simply choosing a menu option by hitting a touchtone on his phone or speaking a word or two. Computer or voice processing software can be organized in two basic ways — menu-driven and non-menu driven. Menu-driven programs are easier for users to use, but they can only present as many options as can be reasonably spoken in a few seconds. Audio menus are typically played to callers in automated attendant/voice messaging, voice response and transaction processing applications. See also MENU and PROMPTS.

AUDIOTEX

A generic term for interactive voice response equipment and services. Audiotex is to voice what on line data processing is to data terminals. The idea is you call a phone number. A machine answers, presenting you with several options, "Push 1 for information on Plays, Push 2 for information on movies, Push 3 for information on Museums." If you push 2, the machine may come back, "Push 1 for movies on the south side of town, Push 2 for movies on the north side of town, etc." See also INFORMATION CENTER MAILBOX.

AUTOMATED ATTENDANT

A device which is connected to a PBX. When a call comes in, this device answers it and says something like, "Thanks for calling the ABC Company. If you know the extension number you'd like, enter that extension now and you'll be transferred. If you don't know it, enter "0" (zero) and the live operator will come on. Or, wait a few seconds and the operator will come on anyway." Sometimes the automated attendant might give you other options, such as, "dial 3" for a directory. Automated attendants are very new. They are connected also to voice mail systems ("I'm not here. Leave a message for me."). Some people react well to automated attendants. Others don't. A good rule to remember is before you spring an automated attendant on your people/customers/subscribers, etc., let them know. Train them a little. Ease them into it. They'll probably react more favorably than if it comes as a complete surprise.

The first impression is rarely forgotten, so try to make it a good experience for the caller. See also DIAL BY NAME.

BACKBONE
The part of a communications network that carries network traffic between access devices.

BAUD
The number of changes in signal state per second in a signal sent by a modem A baud may contain four or more bits. Sometimes confused with BPS, the bits per second transmitted on the channel.

BAUD RATE
The transmission rate of a communications channel. Technically, baud rate refers to the maximum number of changes that can occur per second in the electrical state of a communications circuit. Under the RS-232C communications protocol, 300 baud is likely to equal 300 bps, but at higher baud rates the number of bits per second transmitted is actually higher than the baud rate because one change can represent more than one bit of data. For example, 1,200 bps is usually sent at 600 baud by sending two bits of information with each change in the electrical state of the circuit.

BINARY
A term used to refer to a system of numbers having 2 as its base, such as the digits 0 and 1. A binary system of numbers is used by digital computers because they can only represent data as two states: on or off. (See Digital.)

BISDN
See ISDN.

BIT
Binary digit. The smallest amount of information in a binary system, a 0 or 1 condition.

BPS
Bits per second.

BUFFER
A defined amount of memory assigned to holding data temporarily. This type of information storage is often used to compensate for differences in processing speeds between computers, components, and peripherals, or to free the computer to carry out other operations as when, for example, a complete file is downloaded to a printer's buffer.

BYTE
A group of 8 bits, making up a single memory location. Most computers cannot address a bit, they can only address byte.

BROADCASTING (FAX)
This procedure allows the user to fax a document to a group of people or companies and if desired, personalize each document. Groups can be temporarily or permanently stored in the telephone directories for repeated broadcasting.

BUSY
In use. "Off-hook". There are slow busies and fast busies. Slow busies are when the phone at the other end is busy or off-hook. They happen 60 times a minute. Fast busies (120 times a minute) occur when the network is congested with too many calls. Your distant party may or may not be busy, but you'll never know because you never got that far.

CALL CENTER
A place where call are answered and calls are made. A call center will typically have lots of people (also called agents), an automatic call distributor, a computer for order-entry and look-up on customers' orders. A Call Center could also have a predictive dialer for making lots of calls quickly.

CALL COMPLETION
This is industry jargon for "putting the call through". When a call has been completed, there is an unbroken ("complete") circuit made between the caller and recipient of the call. This circuit is known as the talk path.

CALL PROGRESS ANALYSIS
The automated determination by a piece of telecommunications equipment as to the result of dialing a number. For example, the result of the analysis might be a busy tone, ringing at the other end but no answer after a pre-set number of rings, an answered call and so on. The analysis involves detecting the various call progress tones which will be generated by the telephone network as the call is put through.

CALL PROGRESS MONITORING
Closely analogous to call progress analysis, call progress monitoring may be active during the entire length of a conversation. For example, when a call is placed across a PBX or in a country which does not provide for loop current drop disconnect supervision, it may be necessary for equipment to monitor for a "re-order" or dial tone to determine that the caller hung up. This would be classified as call progress monitoring since it must take place during the entire call, not just when a number is dialled or a transfer is initiated.

CALL PROGRESS TONE
A tone sent from the telephone switch to tell the caller of the progress of the call. Examples of the common ones are dial tone, busy tone, ringback tone, error tone, re-order, etc. Some phone systems provide additional tones, such as confirmation, splash tone, or a reminder tone to indicate that a feature is in use, such as confirmation, hold reminder, hold, intercept tones.

CALLER ID
A name for a service which displays the calling party's telephone number on a special display device.

CALLING TONE
See CNG.

CARRIER FREQUENCY
The frequency of a carrier wave.

CARRIER WAVE

A wave having at least one characteristic that may be varied from a known reference value by modulation.

CAS

Communicating Applications Specification. A high-level API (application programming interface) developed by Intel and DCA that was introduced in 1988. CAS enables software developers to integrate fax capability and high-speed, error-corrected file transfer into their applications.

CBF

Computer-Based Fax.

CCITT

Consultative Committee for International Telephone and Telegraph. On March 1, 1993, the CCITT changed its name to the International Telecommunications Union (ITU-T).

CED

Called Station Identification. A tone used in the hand-shaking used to set up a fax call: the response from a fax machine to the called machines CNG tone.

CENTRAL OFFICE

A telephone company facility where subscriber's lines are joined to switching equipment for connecting other subscribers to each other, locally and long distance.

CENTREX

Centrex is a business telephone service offered by a local telephone company from a local central office. Centrex is basically single line telephone service delivered to individual desks (the same as you get at your house) with features, i.e. "bells and whistles," added. Those "bells and whistles" include intercom, call forwarding, call transfer, toll restrict, least cost routing and call hold (on single line phones).

Centrex is known by many names among operating phone companies, including Centron and Cenpac. Centrex comes in two variations

— CO and CU. CO means the Centrex service is provided by the Central Office. CU means the central office is on the customer's premises.

CFR
Confirmation to Receive frame.

CHANNEL
A path of communication, either electrical or electromagnetic, between two or more points. Also called a circuit, facility, line, link, or path. Typically, what a subscriber rents from the telephone company.

CIG
Calling Subscriber Identification. A frame that gives the caller's telephone number.

CLASS 1
The Class 1 interface is an extension of the EIA/TIA (Electronic Industry Association and the Telecommunications Industry Association) specification for Group 3 fax communication. Class 1 is a series of Hayes AT commands that can be used by software to control fax boards. In Class 1, both the T.30 (the data packet creation and decision-making necessary for call setup) and ECM/BFT (error-correction mode/binary file transfer) are done by the host computer.

CLASS 2
A specification that allows the modem to handle these (Class 1) T.30 functions in hardware. Class 2.0 is a specification that allows the serial modem to handle T.30 functions in hardware, as well as providing ECM.

CLASS 3
Provides the same specifications as Class 2, as well the parameters for the conversion of image file data into ITU-T T.4 compressed image for transmission, and reversion of the conversion on reception.

CLIENT
The requesting program in a distributed computing system. The "Client" send requests to servers across a network and waits for indication from the server that the request is complete.

CNG
Calling Tone. The piercing "whistle" tone (1,100 Hz) of a fax machine to inform the caller that it is ready to receive a transmission.

COMPRESSION
Changing the storage or transmission scheme for information so that less space (fewer bits) are required to represent the same information. Compressing data means that less space is required for storage and less time for the transmission of the same amount of data. Comes in two flavors: lossless compression, where the original information can be reconstructed precisely, and lossy compression, where something close to the original can be reconstructed but some details may differ.

COMPRESSION ALGORITHM
The arithmetic formulae that convert a signal into a smaller bandwidth or fewer bit.

COVER PAGE
The first page of a fax message. It generally includes a header, typically the sender company's logo, the recipient's name and fax telephone number, the sender's fax and voice telephone numbers, the system's date and time, a message, and a footer.

CRP
Command Repeat.

CSI (CSID)
Called Subscriber Identification. An identifier whose coding format contains a number, usually the telephone number from the remote terminal used in fax.

CONNECTED SPEECH
A technical term used to describe speech made of a series of utterances which come in relatively quick succession without co-articulation.

See CO-ARTICULATION. Connected speech is intermediate between discrete speech and continuous speech. Usually applied to the capability of a voice recognizer to recognize words from this type of speech.

CONTINUOUS SPEECH
A technical term used to describe speech made of a series of utterances which come in relatively quick succession with co-articulation. See CO-ARTICULATION. Usually applied to the capability of a voice recognizer to recognize words from this type of speech.

CSI
Called Subscriber Identifier. The "name" of a fax device, transmitted to the fax device at the other end in the course of establishing a fax call. Typically a telephone number and/or company name.

CTI
Computer Telephone Integration. A much better name than "voice processing" for voice processing technology.

DAA
Data Access Arrangement. A device required to hook up Customer Provided Equipment (CPE), usually modem and other data equipment, to the telephone network.

DCN
Disconnect frame. Indicates the fax call is done. The sender transmits it before hanging up; it does not wait for a response.

DCS
Digital Command Signal. Signal sent when the caller is transmitting, which tells the answerer how to receive the fax. Modem speed, image width, image encoding, and page length are all included in this frame.

DECADIC SIGNALING
A fancy way of referring to pulse dialing.

DIAL TONE
The sound you hear when you pick up a telephone. Dial tone is a signal (350 + 440 Hz) from your local telephone company that it is alive and ready to receive the number you dial. If you have a PBX, dial tone will typically be provided by the PBX. Dial tone does not come from God or the telephone instrument on your desk. It comes from the switch to which your phone is connected to.

DIALOGIC
Dialogic Corp, Parsippany, NJ, is one of the leading manufacturers of interactive voice processing equipment and software. They sell equipment through value added resellers, dealers and distributors. Many of their dealers "add value" to the Dialogic components by doing their own specialized software programming, tailoring Dialogic products to particular specialized (and useful) applications.

DID
Direct Inward Dialing. You can dial inside a company directly without going through the attendant. This feature used to be an exclusive feature of Centrex but it can now be provided by virtually all modern PBXs and some modern hybrids. Sometimes spelled DDI, especially in the UK.

DID TRUNKS
Are employed to reduce the number of channels between the PBX and the telephone company central office. DID trunks are one-way trunks. A PBX perceives the DID trunk as one of its single-line telephones and can interpret four-digit dialing.

DIGITAL
A term used to refer to discrete, uniform signals of any kind, not necessarily binary, that do not vary in a continuous manner, as do analog signals. Digital signals are identified by specific values such as on and off, and change instantaneously from one state to another. (See, Binary.)

DIGITAL FACSIMILE
A form of fax in which densities of the original are sampled and quantized as a digital signal for processing, transmission, or storage.

DIGITAL SIGNAL PROCESSOR

A specialized digital microprocessor that performs calculations on digitized signals that were originally analog (e.g., voice) and then sends the results on. Their advantage lies in the programmability of digital microprocessors. DSPs can be used for compression of voice signals to as few as 4,800 bps. DSPs are an integral part of all voice processing systems and facsimile devices.

DIGITAL TRANSMISSION

The use of a binary code to represent information. Analog signals like voice or data, are encoded digitally by sampling the signal many times a second and assigning a number to each sampling. Unlike an analog signal which picks up noise along the way, a digital signal can be reproduced precisely.

DIS

Digital Identification Signal.

DNIS

Dialed Number Identification Service. DNIS is a feature of 800 and 900 lines. Let's say you subscribe to several 800 numbers. You use one line for testing your advertisements on TV stations in Phoenix; another line for testing your advertisements on TV stations in Chicago; and yet another for Milwaukee. Now you get an automatic call distributor and you terminate all the lines in one group on your ACD. You do that because it's cheaper to man and run one group of incoming lines. One queue is more efficient than several small ones, etc. You have all your people answering all the calls. You now need to know which calls are coming from where. So your long distance carrier sends you the call's DNIS — the numbers the person dialed to reach you. Those DNIS digits might come to you in many ways, depending on the technical arrangement you have with your long distance company. In-band or out-of-band. ISDN or data channel, etc. Make sure you understand the difference between DNIS and ANI. DNIS tells you the number your caller called. ANI is the number your caller called from.

DPI
Dots Per Inch. A measure of output device resolution. The number of dots a printer can place in a horizontal inch. The higher the number, the sharper the resolution.

DTMF
Dual Tone Multi-Frequency. A fancy term describing push button or Touchtone dialing. (Touchtone is a registered trademark of AT&T.) In DTMF, when you touch a button on a pushbutton pad, it makes a tone, actually a combination of two tones, one high frequency and one low frequency. Thus the name Dual Tone Multi Frequency. In U.S. telephony, there are actually two types of "tone" signaling, one used on normal business or home pushbutton/touchtone phones, and one used for signaling within the telephone network itself. When you go into a central office, look for the testboard. There you'll see what looks like a standard touchtone pad. Next to the pad there will be a small toggle switch that allows you to choose the sounds the touchtone pad will make — either normal touchtone dialing (DTMF) or the network version (MF).

The eight possible tones that comprise the DTMF signaling system were specially selected to easily pass through the telephone network without attenuation and with minimum interaction with each other. Since these tones fall within the frequency range of the human voice, additional considerations were added to prevent the human voice from inadvertently imitating or "falsing" DTMF signaling digits. One way this was done to break the tones into two groups, a high frequency group and a low frequency group. A valid DTMF tone has only one tone in each group. Here is a table of the DTMF digits with their respective frequencies. One Hertz (abbreviated Hz.) is one cycle per second of frequency.

DTC
Digital Transmit Command.

E-1
The European term for T1. The E1 line bit rate is usually 2.048 Mbps (T1 in the U.S. and Canada is 1.544 Mbps), but variations between the two are not so great that a multiplexer cannot convert

between them. Conversion of E1 to T1 involves both the compression law and signaling format.

ECM

Error Correction Mode. When a fax signal is distorted by a noise pulse induced by electrical interference, or any other reason, errors can occur in the bits transmitted over the telephone line. Without ECM, these errors accumulate and may cause the receiving fax device to disconnect, requiring that the call process be restarted from the beginning, regardless of how much correct material has already been received. ECM provides encapsulated data within HDLC frames, giving the receiver an opportunity to check for, and request retransmission of garbled data.

EDI

Electronic Data Interchange. A series of standards providing automated computer-to-computer exchange of business documents (structured business data, editable documents, or electronic transactions such as invoices, purchase order, etc.) between different companies and computers over telephone lines.

EMAIL

Electronic Mail. A popular application on both LANs and WANs which provides communication among users. There is a variety of systems which vary considerably in their level of sophistication. E-mail services can include simple message handling as well as complex file sharing.

EMail GATEWAYS

All E-mail users can send and receive faxes from within the company's E-mail package. Because the fax messaging is integrated directly into the mail system, fax users have virtually no new software to learn and can take advantage of E-mail features such as workgroup distribution lists, attaching graphics files to text messages, support for different network operating systems, and instant notification of received messages. With an E-mail fax gateway, it is possible to send the same message to both mail users and fax recipients.

ELECTRONIC MAIL INTEGRATION
Integrating computer fax and E-mail technology allows the user to send and receive fax documents using a company's E-mail program. The most popular method uses MHS for Novell networks.

EOM
End Of Message frame. A frame from the sender indicating that the message is done, and that Phase B can be repeated. See, EOP.)

EOP
End of Procedure frame. A frame indicating that the sender wants to end the call.

ETHERNET
A LAN used for connecting computers, printers, workstations, terminals, etc., within the same building. Ethernet operates over twisted pair wire and over coaxial cable at speeds up to 10 Mbps.

FACSIMILE
Facsimile or fax equipment allows information (written, typed, or graphic) to be transmitted through the switched telephone system and printed at the other end. The sending fax scans the material to be sent, digitizing it into binary bits and sending those through a modem to the receiving fax, which essentially reverses the process and outputs it through a printer.

FAX
Abbreviation for facsimile.

FAX ACTIVITY RECORDS
Detailed activity logs that provide the status of tasks: the pending log for tasks still being processed, the received log for incoming faxes; and the sent log!! for outgoing faxes.

FAXBIOS
An "API" used for in-application faxing. Developed by WordPerfect Corp. and Everex Systems, Inc.

FAX BOARD
A specialized synchronous modem designed to transmit and receive facsimile documents. Many also allow for binary synchronous file transfer and V.22 communication.

FAX SERVER
In a LAN, a PC or a self-contained unit that has fax circuitry accessible to all the network's workstations. The server receives requests for fax services and manages them so that they are answered in an orderly, sequential manner. It is used to send and receive faxes by any network user, sharing the common resource of one or more fax boards. Depending on application, a fax server may have a specialized interactive voice response system that routes faxes to a fax machine the user designates by touchtone numbers. The receiving unit may be the user's or one designated by him.

FAX BROADCAST
See Broadcast Fax.

FAX MAIL
Analogous to, and perhaps a feature of, voice mail. Fax mail allows a caller to fax a message rather than speaking a message. Fax messages may be retrieved by the mailbox owner from a fax machine or desktop PC which is able to access the stored file and display it as an image on the computer screen. Some voice mail systems allow fax messages to be incorporated into mailboxes.

FOD
Fax-On-Demand. See Fax-On-Demand.

FAX-ON-DEMAND (FOD)
An enhanced fax technology, advanced Store And Forward. A typical use for fax on demand is to provide product information to potential customers. A caller dials a voice processing unit and selects one or more documents of interest using touchtone menus. If the caller is calling from a fax machine, transmission can being immediately (this is called one-call or *same-call* faxing). If the caller is using a telephone rather than a fax machine, a fax number can be entered

in response to a menu prompt and the fax on demand system will make a later call to that number to deliver the document.

FAX STORE AND FORWARD
This refers to the ability of a computer to store a received fax document as a file stored on a hard drive and re-transmit the document in a subsequent call. Analogous to voice store and forward, which simply means "record" and "play back".

FAX SYNTHESIS
The ability of a computer to create a fax document from stored ASCII text, word processing, database, spreadsheet or other information.

FRAME
A group of data bits in a specific format, with a flag at each end to indicate the beginning and end of the frame. The defined format enables network equipment to recognize the meaning and purpose of specific bit.

FRAME RELAY
Frame relay switching is a form of packet switching, but uses smaller packets and requires less error checking than traditional forms of packet switching. Like traditional X.25 packet networks, frame relay networks use bandwidth only when there is traffic to send. Frame relay does not support voice.

FREQUENCY
The number of complete oscillations per second of an electromagnetic wave.

FTT
Failure-to-train signal.

FULL DUPLEX
A communications protocol in which the communications channel can send and receive signals at the same time.

FSK
Frequency Shift Keying. A modulation technique for translating 1s and 0s into something that can be carried over telephone lines, like sounds. A 1 will be assigned a certain frequency of tone, and a 0 another tone. The transmission of the bits keys the sounds to shift from one frequency to the other.

GPI
A GammaFax Programmer's Interface. C-level programming language, Real-time applications for fax switches and gateways. DOS or OS/2 operating systems.

GROUP 1
Analog fax equipment, according to Recommendation T.2 of the ITU-T. It sends a US letter (8½ by 11") or A4 page in about six minutes over a voice-grade telephone line using frequency modulation with 1.3 KHz corresponding to while at 2.1 KHz to black. North American six-minute equipment uses a different modulation scheme, and is therefore not compatible.

GROUP 2
Analog fax equipment, according to Recommendation T.3 of the ITU-S. It sends a page in about three minutes over a voice grade telephone line using 2.1 KHz AM-PM-VSB modulation.

GROUP 3
A digital fax standard that allows high-speed, reliable transmission over voice grade phone lines. All modern fax devices use Group 3, which is based on ITU-S Recommendation T.4. (Most common worldwide, accounts for more than 90% of all fax machines.)

GROUP 4
A ITU-S fax standard primarily designed to work with ISDN. It is considered difficult to implement, and is not in widespread use owing to the low penetration of ISDN (Group 4 cannot work on non-ISDN lines).
(Group 4's future is questionable.)

HALF DUPLEX
A communications protocol in which the communications channel can handle only one signal at a time. The two stations alternate their transmissions.

HANDSHAKING
An exchange of signals between the fax transmitter and the fax receiver to verify that transmission can proceed, determine which specifications will be used, and to verify reception of the documents sent.

HDLC
High-Level Data-Link Control Standard. It always contains a frame called the Digital Identification Signal (DIS), which describes the standard ITU-T features of the machine. It can also contain two other frames: a Non-Standard Facilities (NSF) frame, which tells the caller about vendor-specific features, and, usually, a Called Subscriber Identification (CSI or CSID) frame, which contains the answerer's telephone number.

HOMOLOGATION
The process of obtaining approval from the local regulatory authorities to attach a device to the public telecommunications network. See also PTT.

HOST
The computer in which a fax board or data modem board resides.

HOT KEY
Refers to TSR utilities in DOS or filters in a Windows environment that allow users to fax without leaving their present application. The ability to send a fax from within an application is one of the most important features of computer fax technology.

HUFFMAN ENCODING
A popular lossless data compression algorithm that replaces frequently occurring data strings with shorter codes. Some implementations include tables that predetermine what codes will be generated from a particular string. Other versions of the algorithm build the code

table from the data stream during processing. Huffman encoding is often used in image compression. (See, Modified Huffman Code.)

HUNT

Refers to the progress of a call reaching a group of lines. The call will try the first line of the group. If that line is busy, it will try the second line, then it will hunt to the third, etc. See also HUNT GROUP.

HUNT GROUP

A series of telephone lines organized in such a way that if the first line is busy the next line is checked ("hunted") and so on until a free line is found. Often this arrangement is used on a group of incoming lines. Hunt groups may start with one trunk and hunt downwards. They may start randomly and hunt in clockwise circles. They may start randomly and hunt in counter-clockwise circles. Inter-Tel uses the terms "Linear, Distributed and Terminal" to refer to different types of hunt groups. In data communications, a hunt group is a set of links which provides a common resource and which is assigned a single hunt group designation. A user requesting that designation may then be connected to any member of the hunt group. Hunt group members may also receive calls by station address.

ICFA

International Computer Facsimile Association. Formed in 1991, its members include the leading companies in the communications and computer industries.

INFORMATION PROVIDER

A business or person providing information to the public for money. The information is typically selected by the caller through touch tones, delivered using voice processing equipment and transmitted over tariffed phone lines, e.g., 900, 976, 970. Typically, billing for information providers' services is done by a local or long distance phone company. Sometimes the revenues for the service are split by the information provider and the phone company. Sometimes the phone company simply bills a per minute or flat charge.

INTERACTIVE VOICE RESPONSE (IVR)

Think of Interactive Voice Response as a voice computer. Where a computer has a keyboard for entering information, an IVR uses remote touchtone telephones. Where a computer has a screen for showing the results, an IVR uses a digitized synthesized voice to "read" the screen to the distant caller. Whatever a computer can do, an IVR can too, from looking up train timetables to moving calls around an automatic call distributor (ACD). The only limitation on an IVR is that you can't present as many alternatives on a phone as you can on a screen. The caller's brain simply won't remember more than a few. With IVR, you have to present the menus in smaller chunks.

Some people use IVR as a synonym for voice processing, or all computer-telephone integration technology involving spoken responses from the computer.

INTERRUPT

The temporary pause of a task caused by an event outside that process. An interrupt signal from hardware, such as a modem in a PC, temporarily suspends other ongoing tasks while the CPU performs the task requested by the interrupting device. Once the routine is completed, the CPU returns to the original tasks.

ISDN

Integrated Services Digital Network. A collection of standards that define interfaces for, and operation of, digital switching equipment developed by carriers, equipment manufacturers, and international standards organizations. It is intended to form the basis for the next generation telephone network and is currently being implemented by carriers throughout the world. Instead of one analog telephone line, there would be two 64 kbps "bearer" lines and one 16 kbps data line. Each bearer line could carry voice, video, data, images or combinations of these. As the name implies, it would be a point-to-point digital system.

ISO

The International Standards Organization. Organization in Paris, devoted to developing standards for international and national data communications. The U.S. representative to the ISO is ANSI.

ITU-T
International Telecommunications Union-Telecommunications. One of four permanent parts of the International Telecommunications Union, based in Geneva, Switzerland. It issues recommendations for standard applying to modem, packet switched interfaces, V.24 connectors, etc. Although it has no power of enforcement, the standards it recommends are generally accepted and adopted by industry. Until March 1, 1993, the ITU-T was known as the CCITT.

LAN
Local Area Network. A short-distance network, within a building or campus, used to link computers and peripheral devices under a form of standard control.

LOCAL LOOP
The wire which passes between your home phone and the phone company. Generally a length of good, old-fashioned copper wire.

LOOP
1. Typically a complete electrical circuit. 2. The loop is also the pair of wires that winds its way from the central office to the telephone set or system at the customer's office, home or factory, i.e. "premises" in telephones. 3. In computer software. A loop repeats a series of instructions many times until some prestated event has happened or until some test has been passed.

LOOP CURRENT DETECTION
When a fax board (or any other modem or telephone) seizes the line (i.e., completes the connection between tip-and-ring terminals of the telephone cable), current flows from the positive battery supply in the telephone central office, through the twisted pair in the loop, through the board, and back to the central office negative terminal where it is detected, showing that this telephone line is off hook. The fax board also detects the loop current and can detect problems

such as disconnects, shutting down the connection or a busy signal, making it wait and redial.

LOSS-LESS COMPRESSION CODING
A coding designed not to lose any data when compressing or restoring an image.

MAILBOX
A set of stored messages belonging to a single owner. Typically, these will be recorded voice messages, but increasingly mailboxes also include E-mail and fax documents.

MAPI
Messaging Application Programming Interface. Developed by Microsoft.

MCF
Message Confirmation Frame. Confirmation by the receiver that it is ready to receive the next page, starting Phase C again.

MINI/MAINFRAME FAX SERVERS
Although many businesses generate and store information on mainframes or minicomputers, most computer-based fax hardware is PC-based. Many mini/mainframe fax servers function in the same way as LAN-fax servers: users can send and receive faxes from their terminals, saving both time and money while they increase the quality of their fax transmissions. By being on the same computer system as an organization's data, mini/mainframe servers can provide an EDI-like component for large, vertical applications such as accounting or purchase order systems, broadening the electronic reach of these programs to any recipient with a fax machine.

MODEM
Acronym for modulator/demodulator. Equipment that converts digital signals to analog signals and vice-versa. Modems are used to send data signals (digital) over the telephone network, which is usually analog. A modem modulates binary signals into tones that can be carried over the telephone network. At the other end, the demodulator part of the modem converts the tones back to binary code.

MODIFIED HUFFMAN CODE (MH)

A one-dimensional data compression technique that compresses data in an horizontal direction only and does not allow transmission of redundant data. Huffman encoding is a lossless data compression algorithm that replaces frequently occurring data strings with shorter codes. Often used in image compression.

MODIFIED READ

Relative Element Address Differentiation code. A two-dimensional compression technique for fax machines that handles the data compression of the vertical line and that concentrates on space between the lines and within given characters.

MODIFIED MODIFIED READ

A two-dimensional coding scheme for Group 4 fax, but now finding use with Group 3 machines.

MPS

Multi-Page Signal. A frame sent if the sender has more pages to transmit.

MVIP

Multi Vendor Integration Protocol. Picture a printed circuit card that fits into an empty slot in a personal computer. The slot carries information to and from the computer. This is called the data bus. Printed circuit cards that do voice processing typically have a second "bus" — the voice bus. That "bus" is actually a ribbon cable which connects one voice processing card to another. The ribbon cable is typically connected to the top of the printed circuit card, while the data bus is at the bottom. As of writing (summer, 1991) there were three "standard" communications buses defined and accepted. Two buses were defined by Dialogic Corporation. They are called The Analog Expansion Bus and the PCM Expansion Bus. The other bus (called MVIP) is from Natural MicroSystems. Both companies have co-opted a number of companies to accept their standard. Both buses can handle voice, data and fax. Here is a write-up on MVIP from Mitel Semiconductor, which has adopted MVIP (as have over 30 other manufacturers):

"The MVIP consists of communications hardware and software that allows printed circuit cards from multiple vendors to exchange information in a standardized digital format. The MVIP bus consists of eight 2 megabyte serial highways and clock signals that are routed from one card to another over a ribbon cable. Each of these highways is partitioned into 32 channels for a total capacity of 256 voice channels on the MVIP bus. These serial link from one card to another. They are electronically compatible with Mitel's ST-BUS specification for inter-chip communications. By letting expansion cards exchange data directly, the MVIP bus opens the PC architecture to voice/data applications that would otherwise overburden the PC processor with data transfers. The MVIP bus is equivalent to an extra backplane that is capable of routing circuit switched data.

"MVIP systems generally have two types of cards; network cards and resource cards. They differ by the switching they provide and in the way they are wired to the bus. Network cards almost always provide more flexible switching and can drive either the input or the output side of the bus, although they usually drive the output side of the bus. Resource cards usually provide very little switching and are only able to drive the input side of the bus. Resource cards usually rely on the network cards to do most or all of the switching on the MVIP bus."

MENU
Options displayed on a computer terminal screen or spoken by a voice processing system. The user can choose what he wants done by simply choosing a menu option — either typing it on the computer keyboard, hitting a touchtone on his phone or speaking a word or two. There are basically two ways of organizing computer or voice processing software—menu-driven and non-menu driven. Menu-driven programs are easier to use but they can only present as many options as can be reasonably crammed on a screen or spoken in a few seconds. Non-menu driven systems may allow more alternatives but are much more complex and frightening. It's the difference between receiving a bland "A" or "C" prompt on the screen — as in MS-DOS and receiving a menu of "Press A if you want Word Processing," "Press B if you want Spread Sheet," etc.

MULTI-TASKING

Doing several different tasks at the same time on one computer. This should not be confused with TASK SWITCHING. In task switching, the computer jumps from one task to another, typically in response to a command from you, the user. For example, in task switching, you might temporarily stop your word processing, jump into your communications package, dial up a database, grab some information, then jump back into your word processor and put that new information into the document you're word processing. However, in true multi-tasking, you could have told your computer to dial the database, grab the information and alert you when it had grabbed the material. At that point you could have included it in your document. But in the meantime, you could have been happily doing word processing.

NAP

Network Applications Platform. (Unisys, Blue Bell, PA, term). A public telephone network service with the ability to send and receive facsimile documents to standard Group 3 fax machines.

NETWORK

An interconnected group of systems. Computer networks connect different and all types of computers and terminals and peripherals, as well as communications systems.

NETWORK FAX SERVER

Allows users to send faxes from their network workstations. Many fax servers allow the faxing of documents from the applications that created them, leaving users free to continue with their work. Incoming faxes can be directed back to your workstation as well. Different fax server products offer different ways of handling incoming faxes, but most use one of the following: the fax can be received by an operator who forwards it to the proper workstation, a fax extension can be assigned to particular workstations for touchtone routing, or there can be a direct fax telephone line accessed through a DID line.

NETWORK INTERFACE MODULE
Electronic circuitry connecting a system (typically a PC) to the telephone network. Network interface modules come in as many versions as there are ways of connecting to the telephone network-from simple loop start telephone lines to complex primary rate interfaces (PRI) on ISDN. Usually, the network interface module slides into one of the expansion slots inside a PC. The board transmits and receives messages from the resource modules providing access to the telephone network.

NETWORK MODULE
Dialogic-speak for a voice processing component which connects to a phone line, whether it is an plain only analog line, digital trunk, DID circuit or whatever.

NODE
A point of connection into a network. In LANs, it is a device on the network. In packet switched networks, it is one of the many packet switches that form the network's backbone.

NOISE
Unwanted electrical signals introduced into telephone lines by circuit components or natural disturbances that tend to degrade the line's performance.

NSC
Non-Standard Facilities Command. A response to the called fax DIS response.

NSF
Non-Standard Facilities frame. Information sent by one fax device to another indicating vendor-specific facilities beyond the standard Group 3 requirements.

NSS
Non-Standard Facilities Setup command, a response to an "NSF" frame.

OCTET
The ITU-T standard term for "byte".

OFF-HOOK
When the handset is lifted from its cradle it's Off-Hook. Lifting the hookswitch alerts the central office that the user wants the phone to do something like dial a call. A dial tone is a sign saying "Give me an order." The term "off-hook" originated when the early handsets were actually suspended from a metal hook on the phone. When the handset is removed from its hook or its cradle (in modern phones), it completes the electrical loop, thus signaling the central office that it wishes dial tone. Some leased line channels work by lifting the handset, signaling the central office at the other end which rings the phone at the other end. Some phones have autodialers in them. Lifting the phone signals the phone to dial that one number. An example is a phone without a dial at an airport, which automatically dials the local taxi company. All this by simply lifting the handset at one end — going "off-hook."

ON-HOOK
When the phone handset is resting in its cradle. The phone is not connected to any particular line. Only the bell is active, i.e. it will ring if a call comes in. See ON-HOOK DIALING and OFF-HOOK.

ON-HOOK DIALING
Allows a caller to dial a call without lifting his handset. After dialing, the caller can listen to the progress of the call over the phone's built-in speaker. When you hear the called person answer, you can pick up the handset and speak or you can talk hands-free in the direction of your phone, if it's a speakerphone. Critical: Many phones have speakers for hands-free listening. Not all phones with speakers are speakerphones — i.e. have microphones, which allow you to speak, also.

ONE-CALL Fax-On-DEMAND
The caller dials an automated service from a fax machine, selects a document through touchtone and then hits the Start button on the fax machine to send or receive a fax. This is "one-call" or "same-

call" fax, as opposed to "two-call" fax where the caller enters a PIN code or fax phone number, the computer then calls back in a later call with the document.

ONE-DIMENSIONAL CODING

A data compression scheme that considers each scan line as being unique, without referencing it to a previous scan line. One-dimensional coding operates horizontally only.

OVERRUN

Loss of data that takes place when the receiving equipment is unable to accept data at the rate it is being transmitted.

OPERATOR INTERCEPT

When an invalid number is dialed or an error condition occurs on the network, an operator intercept may occur. In the US, SIT tones are heard followed by a recorded message explaining the problem.

PACKET

A bundle of data, usually in binary form, organized in a specific way for transmission.

PACKET SWITCHING

Sending data in packets through a network to a remote location. The data are subdivided into individual packets of data, each with a unique identification and individual destination address. This way each packet can take a different route and may arrive in a different order than it was shipped. The packet ID allows the reassembling of data in the proper sequence. This is an efficient way to move digital data. Although it has been used with fax messages, is not yet useful for voice.

PAY-PER-CALL

Some phone calls have an added charge which is levied by the phone company and sent to an information provider. Services such as weather, sports scores, stock prices etc. may be provided. They are generally reached by dialing numbers with a special area code or prefix. In the US, national numbers with the 900 area code or local calls with the 976 prefix have added charges. Similar services are now available in many countries. These are termed "pay-per-call" or "premium rate" services.

PBX

Private Branch eXchange. A private (i.e. you, as against the phone company owns it), branch (meaning it is a small phone company central office), exchange (a central office was originally called a public exchange, or simply an exchange). In other words, a PBX is a small version of the phone company's larger central switching office. A PBX is also called a Private Automatic Branch Exchange, though that has now become an obsolete term. In the very old days, you called the operator to make an external call. Then later someone made a phone system that you simply dialed nine (or another digit — in Europe it's often zero), got a second dial tone and dialed some more digits to dial out, locally or long distance. So, the early name of Private Branch Exchange (which needed an operator) became Private AUTOMATIC Branch Exchange (which didn't need an operator). Now, all PBXs are automatic. And now they're all called PBXs, except overseas where they still have PBXs that are not automatic.

PCM

Pulse Code Modulation. The most common method of encoding an analog voice signal and encoding it into a digital bit stream. First, the amplitude of the voice conversation is sampled. This is called PAM, pulse amplitude modulation. This PAM sample is then encoded (quantized) into a binary (digital) code. This digital code consists of zeros and ones. The voice signal can then be switched, transmitted, and stored digitally.

PEB

PCM Expansion Bus. Dialogic Corp's name for the digital electrical connection between its network interface modules, voice store-and-forward modules, and resource modules. This bus is now open. Technical specifications are available, thus enabling outsiders to create their own resource modules and/or network interface modules.

PEL

Picture Element. Contains only black and white information, no grey shading.

PHASE A, B, C1, C2, D AND E (See below)
The stages in a fax transmission:
>Phase A: Establishment.
>Phase B: Pre-message procedure.
>Phases C1 and C2: In-message procedure, data transmission.
>Phase D: Post-message procedure.
>Phase E: Call release.

PHASE A

In a fax device's call process, the call establishment, or when the transmitting and receiving units connect over the telephone line, recognizing one another as fax machines. This is the start of the handshaking procedure.

PHASE B

In a fax device's call process, the pre-message procedure, where the answering machine identifies itself, describing its capabilities in a burst of digital information packed in frames conforming to the HDLC standard.

PHASE C

In a fax device's call process, the fax transmission portion of the operation. This step consists of two parts "C1" and "C2", which take place simultaneously. Phase C1, deals with synchronization, line monitoring, and problem detection. Phase C2 includes data transmission.

PHASE D

In a fax device's call process, this phase begins once a page has been transmitted. Both the sender and receiver revert to using HDLC packets as during Phase B. If the sender has further pages to transmit, it sends an MPS frame, and the receiver replies with an MCF and Phase C recommences for the following page.

PHASE E

In a fax device's call process, the call release portion. The side that transmitted last sends a DCN frame and hangs up without awaiting a response.

PIN
Procedure Interrupt Negative.

PIXEL
Picture Element. The smallest area of an original, sampled and represented by an electrical signal. A pixel has more than two levels of greyscale information.

PIP
Procedure Interrupt Positive.

POLLING
Refers to some form of data or fax network arrangement whereby a central computer or fax machine/board very quickly asks each remote location in turn whether they want to send some information. The purpose is to give each user or remote data terminal an opportunity to transmit and receive information on a circuit or using facilities which are being shared. Polling is typically used in a multipoint or multidrop line. It is done mostly to save money on telephone lines.

PPO-UP
See HOT KEY

PORT
An entrance to or exit from a network, the physical or electrical interface through which one gains access. The interface between a process or a program and a communications or transmission facility. A point in the computer or telephone system where data may be accessed. Peripherals are connected to ports.

PORT DENSITY
Jargon for "the number of ports, ie. phone lines, supported by a system" or "the number of ports per voice processing board". "High density" means that a lot of ports are handled by one or a few boards.

POSTSCRIPT
A page description language for PCs, which is standard for many graphics and desktop publishing applications. It allows the creation

of elaborate documents. Other applications such as spreadsheets, word processors, and databases rarely require this level of sophistication.

POTS
Plain Old Telephone Service. The basic service supplying standard single line telephones, telephone lines and access to the public switched network. Nothing fancy. No added features. Just receive and place calls. Nothing like Call Waiting or Call Forwarding. They are not POTS services. Pronounced POTS, like in pots and pans.

PRI-EOM
Procedure Interrupt-End Of Message.

PRI-EOP
Procedure Interrupt-End Of Procedures.

PRI-MPS
Procedure Interrupt-Multipage Signal.

PRINTER EMULATION
Enables the mimicking of a printer-generated document. This way, the outgoing fax will look as if it has come from the printer attached to a computer. This can include full formatting, as well as letterhead, signature, and different graphic images.

PRIVATE BRANCH EXCHANGE
A business phone system, often abbreviated to PBX or PABX ("A" for Automatic).

PROTOCOL
A specified set of rules, procedures, or conventions, relating to format and timing of data transmission between two devices. A standard procedure that two data devices must accept and use to be able to understand each other.

PSK
Phase Shift Keying. A method of modulating the phase of a signal to carry information.

PSTN
See PUBLIC SWITCHED TELEPHONE NETWORK.

PTT
The Post Telephone and Telegraph (PTT) administrations, usually controlled by their governments, provide telephone and telecommunications services in most countries where these services are not privately owned. In CCITT documents, these are the entities referred to as "Operating Administrations."

It is not a simple thing to obtain approval from the PTTs to sell and use telecommunication equipment of any kind in their countries. The world is far from being one in the field of telecommunications. Meeting international requirements typically means providing hardware and software modifications to the product, unique to each country, and then going through an extremely rigorous approval process that can average between six to nine months. Products are required to meet both safety and compatibility requirements.

PUBLIC SWITCHED TELEPHONE NETWORK
Usually refers to the worldwide voice telephone network accessible to all those with telephones and access privileges (i.e. In the U.S., it was formerly called the Bell System network or the AT&T long distance network).

PULSE DIALING
One or two types of dialing that uses rotary pulses to generate the telephone number.

PULSE TO TONE CONVERTER
A device which recognizes the "clicks" made by a rotary dial phone and converts them to DTMF tones. Not always a reliable technology. Consider the problem faced by the device in distinguishing the click made by dialing a "1" digit and a static click caused by lightning or other interference on the line.

QAM
Quadrature Amplitude Modulation. A modulation technique that uses variations in signal amplitude, allowing data-encoded symbols

to be represented as any of 16 or 32 different states. Some QAM modems allow dial-up data rates of up to 9,600 bps.

QUEUING
The act of "stacking" or holding calls to be handled by a specific person, trunk, or trunk group.

RAM
Random Access Memory. The primary memory in a computer. Memory that can be overwritten with new information. The "random access" part of the name comes from the fact that the next bit of information in RAM can be located no matter where it is, in an equal amount of time, making access to it considerably faster than to information in other storage media, such as a hard disk.

RASTER SCANNING
The method of scanning in which the scanning spot moves along a network of parallel lines, either from side to side or top to bottom.

REFERENCE LINE
The first scanning line in memory. The location of each black pixel of this line is kept in memory for the next scanned line. Depending on the compression technique used, more or fewer scan lines are necessary.

RESOLUTION
A measure of capability to delineate picture detail.

RESOLUTION LEVELS
Provides for high- and low-resolution levels, (the ITU-TSS Group 3 standard is most popular) up to 400 dots per inch. Most computer fax products allow users to select which resolution meets their needs.

RESOURCE MODULE
A Dialogic term referring to devices which perform specific voice processing functions such as voice compression, voice recognition, facsimile transmission and reception, and conversion of computer text to spoken words over the telephone. Resource modules are

typically connected to the telephone network through network interface modules.

RING

1. As in Tip and Ring. One of the two wires (the two are Tip and Ring) needed to set up a telephone connection. 2. Also a reference to the ringing of the telephone set. 3. The design of a Local Area Network (LAN) in which the wiring loops from one workstation to another, forming a circle (thus, the term "ring"). In a ring LAN, data is sent from workstation to workstation around the loop in the same direction. Each workstation (which is usually a PC) acts as a repeater by re-sending messages to the next PC in the ring. The more PC's, the slower the LAN. Network control is distributed in a ring network. Since the message passes through each PC, loss of one PC may disable the entire network. However, most ring LANs recover very quickly should one PC die or be turned off. If it dies, you can remove it physically from the network. If it's off, the network senses that and the token ignores that machine. In some token LANs, the LAN will close around a dead workstation and join the two workstations on either side together. If you lose the PC doing the control functions, another PC will jump in and take over. This is how the IBM Token-Passing Ring works.

RJ-11

RJ-11 is a six conductor modular jack that is typically wired for four conductors (i.e. four wires). Occasionally it is wired for only two conductors — especially if you're only wiring up for tip and ring. The RJ-11 jack (also called plug) is the most common telephone jack in the world. The RJ-11 is typically used for connecting telephone instruments, modems and fax machines to a female RJ-22 jack on the wall or in the floor. That jack in turn is connected to twisted wire coming in from "the network" — which might be a PBX or the local telephone company central office. RJ-22 wiring is typically flat. None of its conductors (i.e. wires) are twisted. You cannot use flat cable for high-speed data communications, like local area networks. See also RJ-22 and RJ-45.

RJ-14

A jack that looks and is exactly like the standard RJ-11 that you see on every single line telephone. Whereas the RJ-11 defines one line — with the two center, red and green, conductors being tip and ring, the RJ-14 defines two phone lines. One of the lines is the "normal" RJ-11 line — the red and green center conductors. The second line

ROUTING
The process of selecting the correct path for a message.

RS-232
A set of standards, specifying the various electrical and mechanical characteristics for interfaces between computers, terminals, and modems. It applies to synchronous and asynchronous binary data transmission.

RTN
Retrain Negative.

RTP
Retrain Positive.

SAME-CALL FAX
See ONE-CALL FAX.

SCANNER
A device used to input graphic images in digital form. A fax machine's scanner determines the brightness of a document's pixel for transmission.

SCHEDULED TRANSMISSION
A feature allowing the user to schedule a fax transmission at a specific date or time in the future. Key benefits are convenience and cost savings. Scheduling jobs at a period of low telephone rates can have immediate considerable savings.

SC BUS
The next generation bus now under development by Dialogic and others. See SCSA. Will play a role analogous to the PEB.

SCRIPT
1. A written document specifying the wording of menus and informational messages to be recorded when designing a voice response application.
2. The flow-chart or other description specifying the way that a voice response system interacts with a caller.

SCSA
(pronounced "scuzza") Signal Computing System Architecture. SCSA is a standard for all levels of design of voice processing components from the voice bus chip level to the applications programming interface (API).

SCSA is an open standard. A consortium of leading telecommunications and computing technology players, led by Dialogic and including companies such as IBM and Seimens are, at the time of writing, cooperating in development of the SCSA specification. The specification documents will be available to anyone who wants them for a nominal fee. There will be no technology license fees charged to developers who wish to create SCSA-compatible products.

The most important components of SCSA for the voice processing developer are a bus and a uniform API.

The primary bus is the SCbus. SCbus is a high capacity bus which is designed to be the "next generation PEB". Where the PEB has up to 32 time-slots, the SCbus will have up to 2,048 time-slots: enough band-width for high-fidelity audio, full-motion video and other demanding applications of the future. The standardized API should make SCSA components such as voice processing, voice recognition, speech synthesis, video boards and others accessible independent of the component manufacturer.

SERVER
A computer providing a service such as shared access to a file system, a printer or an E-mail system to LAN users. Usually a combination of hardware and software. A process in a distributed computing system that provides a service in response to requests from clients.

SDLC
Synchronous Data-Link Control.

SPOOLER
A program that controls spooling. Spooling, a term mostly associated with printers, stands for Simultaneous Peripheral Operations On Line. Spooling, temporarily stores programs or program outputs on magnetic tape, or disks for output or processing. On a LAN, a printer is controlled by a spooler. The spooler places each print request on the LAN in the print queue and prints it when it reaches the top of the queue.

STANDARDS
Agreed principles of protocol. Standards are set by committees working under various trade and international organizations. RS standards, such as RS-232-C are set by the "EIA", the Electronics Industries Association. "ANSI" standards for data communications are from the X committee. Standards from ANSI would look like X3.4-1967, which is the standard for the ASCII code. The ITU-T does not put out standards, but rather publishes recommendations, owing to the international personalities and countries involved. "V" series recommendations refer to data transmission over the telephone network, while "X" series recommendations, such as X.25, refer to data transmission over public data networks.

SWITCHED NETWORK
A network providing switched communications service; that is, the network is shared among many users, any of whom may establish communication between desired points when required.

SWITCH
A mechanical, electrical or electronic device which opens or closes circuits, completes or breaks an electrical path, or selects paths or circuits.

T-1
Also spelled T1. A digital transmission link with a capacity of 1.544 Mbps (1,544,000 bits per second). T-1 uses two pairs of normal twisted wires, the same as you'd find in your house. T-1 normally

can handle 24 voice conversations, each one digitized at 64 Kbps. T-1 is a standard for digital transmission in North America. It is usually provided by the phone company and used for connecting networks across remote distances. Bridges and routers are used to connect LANs over T-1 networks. In Europe the similar but incompatible service is called E-1 or E1. See A & B BITS for details of signaling.

For a full explanation of T1 see Bill Flanagan's book The Guide to T-1 Networking. (Call 1-800-LIBRARY for your copy.)

T-2
A digital transmission link digital transmission link with a capacity of 6.312 Mbps. T-2 can handle at least 96 voice messages.

T-3
A digital transmission link with a capacity of 44.736 Mbps. T-3 can handle 672 voice messages simultaneously. T-3 runs on fiber optic cable.

T.4
ITU-T recommendation for Group 3 devices, providing definitions of various V-series protocols and signals used during Group 3 operations, including: all supported resolutions, one-dimensional encoding and two-dimensional encoding, and optional error control and error-limiting modes.

T.6
ITU-T recommendation for Group 4 machines. It defines the facsimile coding schemes and their associated coding control functions for black and white images.

T.30
ITU-T recommendation. This handshake protocol describes the overall procedure for establishing and managing communication between two fax devices. It covers five phases of operation: call setup, pre-message procedure (selecting the communication mode), message transmission (including both phasing and synchronization), post-message procedure (EOM and confirmation), and call release.

T.35
ITU-T recommendation proposing a procedure for the allocation of ITU-T members' country or area code for non-standard facilities in telematic services.

T.611
Also known as Appli/COM. A messaging standard proposed by France and Germany, defining a programmable communication interface (PCI) for Group 3 fax, Group 4 fax, teletext, and telex service. It provides communications between the local application and the communications application. This local application to communications application relationship (client/server) exists in several places in the protocol stack.

TCF
Training Check Frame. Last step in a series of signals called a training sequence, designed to let the receiver adjust to line conditions.

TELEFAX
European term for fax.

TELECOPIER
European term for fax.

TEXT-TO-SPEECH
Technology for converting speech in the form of ASCII or other text to a synthesized voice.

TIFF FILE FORMAT
Tagged Image File Format. TIFF provides a way of storing and exchanging digital image data. Aldus Corp., Microsoft Corp., and major scanner vendors developed TIFF to help link scanned images with the popular desktop publishing applications. It is now used for many different types of software applications ranging from medical imagery to fax modem data transfers, CAD programs, and 3D graphic packages. The current TIFF specification supports three main types of image data: Black and white data, halftones or dithered data, and grayscale data. A special variant of TIFF, called TIFF/F, has been defined specifically for storing fax images. Note that most

standard PC graphics software, at least at the time of writing, doesn't support TIFF/F even it does support other flavors of TIFF file.

TIP
1. The first wire in a pair of wires. The second wire is called the "ring" wire. 2. A conductor in a telephone cable pair which is usually connected to positive side of a battery at the telephone company's central office. It is the phone industry's equivalent of Ground in a normal electrical circuit.

TONE DIAL
A pushbutton telephone dial that makes a different sound (in fact, a combination of two tones) for each number pushed. The correct name for tone dial is "Dual Tone MultiFrequency" (DTMF). This is because each button generates two tones, one from a "high" group of frequencies — 1209, 1136, 1477 and 1633 Hz — and one from a "low" group of frequencies — 697, 770, 852 and 841 Hz. The frequencies and the keyboard, layout have been internationally standardized, but the tolerances on individual frequencies do vary between countries. This makes it more difficult to take a touch-tone phone overseas than a rotary phone.

You can "dial" a number faster on a tone dial than on a rotary dial, but you make more mistakes on a tone dial and have to redial more often. Some people actually find rotary dials to be, on average, faster for them. The design of all tone dials is stupid. Deliberately so. They were deliberately designed to be the exact opposite (i.e. upside down) of the standard calculator pad, now incorporated into virtually all computer keyboards. The reason for the dumb phone design was to slow the user's dialing down to the speed Bell central offices of early touch tone vintage could take. Today, central offices can accept tone dialing at high speed. But sadly, no one in North America makes a phone with a sensible, calculator pad or computer keyboard dial. On some telephone/computer work-stations you can dial using the calculator pad on the keyboard. This is a breakthrough. It a lot faster to use this pad. The keys are larger, more sensibly laid out and can actually be touch-typed (like on a keyboard.) Nobody, but nobody can "touch-type" a conventional telephone tone pad. A tone dial on a telephone can provide access to various special services and features — from ordering your groceries over the phone to inquiring into the prices of your (hopefully) rising stocks.

TOUCH TONE
A former trademark once owned by AT&T for a tone used to dial numbers. For a full explanation of touchtone, see DTMF.

TRAINING SEQUENCE
Part of the hand-shake used in establishing a fax call where the two devices can adjust to prevailing line conditions.

TRELLIS CODING
A method of forward error correction used in certain high-speed modems where each signal element is assigned a coded binary value representing that element's phase and amplitude. It allows the receiving modem to determine, based on the value of the preceding signal, whether or not a given signal element is received in error.

TRELLIS CODING MODULATION (TCM)
A version of quadrature amplitude modulation that enables relatively high bit rate signals to be used on ordinary voice-grade circuits. Used as a modem modulation technique in which algorithms are used to predict the best fit between the incoming signal and a large set of possible combinations of amplitude and phase changes. TCM provides for transmission speeds of 14.4 kbps and above on single voice-grade telephone lines.

TRUNK
A communication line between two switching systems. The term "switching system" typically includes the equipment in a telephone company's central office and PBXs. A "tie trunk" connects PBXs. Central office trunks connect a PBX to the switching system at the central office.

TSI
Transmitting Subscriber Information. A frame that may be sent by the caller, with the caller's telephone number (may be used to screen calls).

TSR

Terminate and Stay Resident. A term for loading a software program in a DOS computer in which the program loads into RAM and is always ready for running at the touch of a combination of keys.

TSS

See CCITT.

TTI

Transmit Terminal Identification. The telephone number and words on top of a received fax document, identifying its point of origin. This information does not originate with the telephone company, but with the sender, who programs it into the fax machine.

TWISTED PAIR

Two insulated copper wires twisted around each other to reduce induction (thus interference) from one wire to the other. The twists, or lays, are varied in length to reduce the potential for signal interference between pairs. Several sets of twisted pair wires may be enclosed in a single cable. In cables greater than 25 pairs, the twisted pairs are grouped and bound together in a common cable sheath. Twisted pair cable is the most common type of transmission media. It is the normal cabling from a central office to your home or office, or from your PBX to your office phone. Twisted pair wiring comes in various thicknesses. As a general rule, the thicker the cable is, the better the quality of the conversation and the longer cable can be and still get acceptable conversation quality. However, the thicker it is, the more it costs.

TWO-DIMENSIONAL CODING

A data compression scheme that uses the previous scan line as a reference when scanning a subsequent line. Because an image has a high degree of correlation vertically as well as horizontally, 2-D coding schemes work only with variable increments between one line and the next, permitting higher data compression.

V.17
ITU-T recommendation for simplex modulation technique for use in extended Group 3 facsimile applications only. Provides 7,200, 9,600, 12,000, and 14,400 bps trellis-coded modulation.

V.21
ITU-T recommendation for 300 bps duplex modems for use on the switched telephone network. V.21 modulation is used in a half-duplex mode for Group 3 fax negotiation and control procedures.

V.22
ITU-T recommendation for 1,200 bps duplex modem for use on the switched telephone network and on leased lines.

V.22bis
ITU-T recommendation for 2,400 bps duplex modems for use on the switched telephone network. V.22 also provides for 1,200 bps operation for V.22 compatibility.

V.27ter
ITU-T recommendation for 2.4/4.8-kbps/s modem for use on the switched telephone network. Half-duplex only. It defines the modulation scheme for Group 3 facsimile for image transfer at 2,400 and 4,800 bps.

V.29
ITU-T recommendation for 9,600 bps modem for use on point-to-point leased circuits. This is the modulation technique used in Group 3 fax for image transfer at 7,200 and 9,600 bps. V.29 uses a carrier frequency of 1,700 Hz which is varied in both phase and amplitude. V.29 can be full duplex on four-wire leased circuits, or half duplex on two-wire and dial-up circuits.

V.32
ITU-T recommendation for 9,600 bps two-wire full duplex modem operating on regular dial-up lines or two-wire leased lines. V.32 also provides fallback operation at 4,800 bps.

V.32bis
ITU-T recommendation for full-duplex transmission on two-wire leased and dial-up lines at 4,800, 7,200, 9,600, 12,000, and 14,400 bps. Provides backward compatibility with V.32. It includes a rapid change renegotiation feature for quick and smooth rate changes when line conditions change.

V.33
ITU-T recommendation for 14.4 kbps and 1.2 kbps modem for use on four-wire leased lines.

V.42
ITU-T recommendation, primarily concerned with error correction and compression modems. The protocol is designed to detect errors in transmission and recover with a retransmission.

V.fast
A modem standard under development. Once approved by the ITU-T, it will raise modem speeds to 19.2 kbps.

VOICE BOARD
Also called a voice card or speech card. A Voice Board is a computer add-in card which can perform voice processing functions. A voice board has several important characteristics: It has a computer bus connection. It has a telephone line interface. It typically has a voice bus connection. And it supports one of several operating systems, e.g. MS-DOS, UNIX. At a minimum, a voice board will usually include support for going on and off-hook (answering, initiating and terminating a call); notification of call termination (hang-up detection); sending flash hook; and dialing digits (touchtone and rotary). See VRU.

VOICE INTEGRATION
Allows computer fax solutions to be store-and-forward hubs for both image as well as voice communication. Many of these products work on PC-based systems and offer all the capabilities of a message center.

VOICE MAIL
You call a number. A machine answers. "Sorry. I'm not in. Leave me a message and I'll call you back." It could be a $50 answering machine. Or it could be a $200,000 voice mail system. The primary purpose is the same — to leave someone a message. After that, the differences become profound. A voice mail system lets you handle a voice message as you would a paper message. You can copy it, store it, send it to one or many people, with or without your own comments. When voice mail helps business, it has enormous benefits. When it's abused — such as when people "hide" behind it and never return their messages — it's useless. Some people hate voice mail. Some people love it. It's clearly here to stay.

VOICE MESSAGING
Recording, storing, playing back and distributing phone messages. New York Telephone has an interesting way of looking at voice messaging. NYTel sees it as four distinct areas: 1. Voice Mail, where messages can be retrieved and played back at any time from a user's "voice mailbox"; 2. Call Answering, which routes calls made to a busy/no answer extension into a voice mailbox; 3. Call Processing, which lets callers route themselves among destinations via their touch-tone phones; and 4. Information Mailbox, which stores general recorded information for callers to hear.

VOICE RECOGNITION
The ability of a machine to understand human speech. When applied to telephony environments, the limited bandwidth (range of frequencies transmitted by a telephone connection) and other factors such as background noise and the poor quality of most telephone microphones severely limits the ability of current technology to recognize spoken words. Typical systems are able to recognize standard vocabularies of 16 or so words, such as the digits, yes, no and stop.

VOICE RESPONSE UNIT (VRU
Think of a Voice Response Unit (also called Interactive Voice Response Unit) as a voice computer. Where a computer has a keyboard for entering information, an IVR uses remote touchtone telephones. Where a computer has a screen for showing the results, an IVR uses a digitized synthesized voice to "read" the screen to the distant

caller. An IVR can do whatever a computer can, from looking up train timetables to moving calls around an automatic call distributor (ACD). The only limitation on an IVR is that you can't present as many alternatives on a phone as you can on a screen. The caller's brain simply won't remember more than a few. With IVR, you have to present the menus in smaller chunks. See VOICE BOARD.

VRU
See VOICE RESPONSE UNIT.

WAN
Wide Area Network. A network using common carrier-provided lines that cover an extended geographical area. WANs are data networks typically extending a LAN outside the building, over telephone common carrier lines to link to other LANs in remote buildings or other geographic areas.

WINK
A signal sent between two telecommunications devices as part of a "hand-shaking" protocol. On a digital connection such as a T-1 circuit, a wink is signaled by a brief change in the A and B signaling bits from off to on and back to off (the reverse of a flash-hook). On an analog line, a wink is signaled by a change in polarity (electrical + and -) on the line.

WINK OPERATION
A timed, off-hook signal, normally of 140 ms, which indicates the availability of an incoming register for receiving digital information from the calling office.

WINK START
Short duration off-hook signal. (See Wink Operation.)

WYSIWYG
(Pronounced: Wiz-E-Wig) What You See Is What You Get. You could argue this has nothing to do with fax (it's remote), but I like it anyway.

X.25

Possibly one of the most important int'l standards ever recommended by the ITU-T. From the beginning, it has provided a common reference point by which mainframe computers, word processors, mini-computers, VDUs, microcomputers, and varied equipment from different manufacturers can operate together over a packet switched network. X.25, defines the interface between a public data network and a packet-mode user device. "X" stands for "packet switched network."

X.38

ITU-T recommendation for access of Group 3 facsimile equipment to the Facsimile Packet Assembly/Disassembly (FPAD) facility in public data networks situated in the same country.

X.39

ITU-T recommendation for the exchange of control information and user data between a Facsimile Packet Assembly/Disassembly (FPAD) facility and a packet mode data terminal equipment (DTE) or another pad, for international internetworking.

X.400

ITU-T recommendations for the transmission of electronic text and graphic mail between unlike computers, terminals, and computer networks. Briefly, X.400 describes how mail messages are encoded, and X.25 sets up how they are transmitted.

X.500

The ITU-T standard defining directory services, most commonly for email systems. It provides for common naming and addressing in the networks of large-scale enterprises.

ZERO FILL

A traditional fax device is mechanical. It must reset its printer and advance the page as it prints each scan line it receives. If the receiving machine's printing capability is slower than the transmitting device's data sending capability, the transmitting device adds "fill bits" to pad out the span of send time, giving the slower remote machine the additional time it needs to reset prior to receiving the next scan line.

FAX RESOURCE GUIDE

Brooktrout Technology, Inc. (Universal Voice/Fax Boards)
144 Gould Street
Needham, MA 02194
617-449-4100
617-449-9009 Fax
800-333-5724 FOD

Dialogic Corporation (Voice Boards)
1515 Route 10
Parsippany, NJ 07054
201-993-3000
201-993-3093 Fax
800-755-5599 FOD

Flatiron Publishing, Inc (Books and Magazines on)
12 West 21st Street
New York, NY 10010
212-691-8215
212-691-1191 Fax
2120645-1478 FOD (One-Call)

GammaLink (Fax Boards)
Division of Dialogic Corp
1314 Chesapeake Terrace
Sunnyvale, CA 94089
408-744-1400
408-744-1900 Fax

Ibex Technologies, Inc (Fax-On-Demand Software)
550 Main Street, Suite G
Placerville, CA 95667
916-621-4342
916-621-2004 Fax
800-289-9998 FOD

Instant Information Inc. (Fax Service Bureau)
5 Broad Street
Boston, MA 02110
617-523-7636
617-723-6522 Fax

Parity Software Development Corp (VOS Programming Language)
870 Market Street, Suite 800
San Francisco, CA 94102
415-989-0330
415-989-0441 Fax

SpectraFAX Corp (Special Request Fax-On-Demand Products)
3050 N. Horseshoe Drive Suite 100
Naples, FL 33942
813-643-5060
813-643-5070 Fax
800-833-1329 FOD

The Kauffman Group (Enhanced Fax Consulting)
324 Windsor Drive
Cherry Hill, NJ 08002
609-482-8288
609-482-8940 Fax

PUBLICATIONS and SUGGESTED READING

Teleconnect Magazine

Computer Telephony Magazine

PC-Based Voice Processing (636 pages, 2nd edition)
by: Bob Edgar, founder, Parity Software Development Company

Telephony for Computer Professionals (248 pages)
by: Jane Laino, President, Corporate Communications Consultants

Newton's Telecom Dictionary (1,179 pages, 7th edition)
by: Harry Newton, Publisher, Flatiron Publishing

Flatiron Publishing, Inc
12 West 21st Street
New York, NY 10010
212-691-8215
212-691-1191 Fax
212-645-1478 FOD (one-call)

Call for a copy of their most recent catalog.

Don't forget: **Computer Telephony '95** Conference and Exposition
(Formerly: Telecom Developers)

March 7-9, 1995
Dallas Convention Center

See you in Dallas!!

FAX-ON-DEMAND DEMONSTRATIONS

Aldus	206-628-5737
AT&T'S World News Faxline	800-637-6306
ATI Technologies	905-882-2600
Compaq Computer	800-345-1518
Dialogic Corp	800-755-5599
Dow Jones JournalFax	800-759-9966
Forbes Magazine	800-274-2793
Fortune Magazine	800-759-5294
Hewlett Packard	208-344-4809 **
IBM	800-426-4329
INC. Magazine	800-995-4455
Lotus Development	617-253-9150
Novell, Inc	800-638-9273
Symantec Corp	800-554-4403
The Economics Press	800-425-1010 x590
The Export Hotline	800-USA-XPORT
Upside Magazine	800-800-8257
US Department of Agriculture	202-690-3944 **

** One-Call Applications - You must be calling from the telephone on your fax machine to access this service.

Note: Depending on the application, document code numbers and extensions may change.

THE KAUFFMAN GROUP

SPECIAL OFFER

Want **FREE** INFORMATION and UPDATES?
Want to learn more?
Want to hear about new fax applications?
Want to receive periodic reports new information?

Be Added to my Fax Broadcast and Mailing Lists!

Send your name, address, phone and **Fax** to:

Maury Kauffman
Managing Partner
The Kauffman Group
324 Windsor Drive
Cherry Hill, NJ 08002

609-482-8288
609-482-8940 Fax

OR

SIMPLY, Fax your business card to: 609-482-8940 and write
"Mailing List," (no cover page necessary.)

I'll do the rest.

Got a newsworthy: Product
Service
Application....   fax me the details!

Other Books We Publish

Flatiron Publishing publishes books and magazines and organizes trade conferences on computer telephony, telecommunications, networking and voice processing. We also distribute the books of other publishers, making us the "central source" for all the above materials. Call or write for your FREE book catalog. These are the titles of some books we publish.

ATM Users' Guide
City, County, State Guide To Acquiring A Phone System
Client Server Computer Telephony
Complete Traffic Engineering Handbook
Computer Based Fax Processing
Customer Service Over the Phone
Frames, Packets and Cells in Broadband Networking
The Guide to Frame Relay
The Guide to SONET
The Guide to T-1 Networking
Local & Long Distance Telephone Billing Practices
Moore's Imaging Dictionary
Newton's Telecom Dictionary
PC-Based Voice Processing
SCSA
Speech Recognition
Telephony for Computer Professionals
VideoConferencing: The Whole Picture
Which Phone System Should I Buy?

Quantity Purchases

If you wish to purchase this book, or any others, in quantity, please contact:

Christine Kern, Manager
Flatiron Publishing, Inc.
12 West 21 Street
New York, NY 10010

212-691-8215 1-800-LIBRARY
facsimile orders: 212-691-1191